TACTICAL NEUTRALIZATION TECHNIQUES

GREGORY J. CONNOR   MATTHEW D. SUMMERS

TACTICAL NEUTRALIZATION TECHNIQUES

GREGORY J. CONNOR　　　　**MATTHEW D. SUMMERS**

ISBN 0-87563-327-7

Copyright © 1988
STIPES PUBLISHING COMPANY

Published by
STIPES PUBLISHING COMPANY
10-12 Chester Street
Champaign, Illinois 61820

TABLE OF CONTENTS

Dedication ii
Foreword iii
Preface iv
About the Authors vii

		Page
Introduction	Use of Force: Confrontation/Control Continuum	1
Chapter 1	Why Physical Training and Personal Defense	5
Chapter 2	Status of "Functional Fitness"	11
Chapter 3	Preparing to Exercise	13
Chapter 4	Self-Assessment—Evaluation Procedures	15
Chapter 5	Exercise Routines (Calisthenics)	25
Chapter 6	Utilization of Personal Weapons	41
Chapter 7	Confrontations: Mental Perspectives	49
Chapter 8	Street Stances	53
Chapter 9	Punching, Kicking, Blocking, Falling Drills	59
Chapter 10	Principles of Escape	87
Chapter 11	Standard Search Technique	115
Chapter 12	Nomenclature and Basic Procedures of Handcuffing	127
Chapter 13	Specialized Search Techniques	141
Chapter 14	Specialized Handcuffing Techniques	153
Chapter 15	Takedown/Handcuffing Combinations	189
Chapter 16	Driver Removal Techniques	215
Chapter 17	Weaponless Comealongs	223
Chapter 18	Escorting and Transporting Procedures	231
Chapter 19	Baton Techniques	239
Chapter 20	Specialized Baton Techniques	257
Chapter 21	Weapon Retention Techniques	267
Chapter 22	Down Fighting Techniques	277
Chapter 23	Nerve Compression Control Techniques	285
Chapter 24	Individualized Chemical Irritants	299

DEDICATION

To my children, Brian, Aaron and Lynn, so that if they seek my source of occupational satisfaction, it will be in an environment of increased safety and enhanced understanding.

G. J. C.

To my family, friends and colleagues, who have provided support, guidance and motivation in this effort.

M. D. S.

FOREWORD

The authors and publisher accept no liability whatsoever for any injuries to persons or property resulting from the application or adoption of any of the procedures presented or implied in this manual.

The concepts and tactics discussed and demonstrated in this manual are those of the authors acting in their individual capacities, and not as members of the faculty of the Police Training Institute, University of Illinois.

These concepts and tactics may not be consistent with those recommended and/or taught by the Police Training Institute, University of Illinois, its faculty or staff, and may not represent the criteria currently used by the Police Training Institute for the Tactical Neutralization Techniques Program.

Preface

THE SYSTEM AND SYMBOL OF TACTICAL NEUTRALIZATION

This manual has been designed to provide the instructional basis essential for initial understanding of Tactical Neutralization Techniques. Each of the procedures outlined are under constant field testing to assure their applicability, and therefore, represent the most effective tactics attuned to contemporary attitudes of those we serve, the intensity and variety of the training mode, and the attitudinal, intellectual, and motivational profile of today's police officer.

The overriding feature throughout our program is that of neutralization, which in action terms connotes a process by which the most acceptable, efficient, and effective "street" techniques are employed by the trained officer to nullify illegal, aggressive activities by the offender. Once even the threshold of neutralization is achieved, the officer's actions should be directed to negate practically based potential resumption of illegal aggression by the subject. At that point, and throughout the entire confrontation, the officer must remain aware of his controlled presence with the offender, ever cognizant of his safety and his role in the continued diffusion and/or redirection of violence.

It is our goal to provide this text and other ancillary materials for the individual officer engaging in a self-paced program of control tactics, or as a structured text for a classroom training design in an academy setting. It should be noted that in the opinion of the authors, maximized training outcomes can best be achieved in a formal, inter-active training environment, under the direction of a certified Tactical Neutralization Techniques instructor.

Indicative of the officer's understanding of concepts, assimilation, and demonstration of techniques presented, and the knowledgeable expression of the philosophy of tactical neutralization, each participant can be evaluated throughout the program. An opinional evaluation package is available and identifies each technique in a functional format of performance. In addition to potential written certification of skill attainment, he can be awarded a certificate designating skill status. This certificate can be provided not only as an indication of earned proficiency but also to provide a personal, visual incentive to the officer.

Additionally, we have presented the philosophy of Tactical Neutralization symbolically via the outlined graphic of the covered fist. The symbology of the covered fist that characterizes Tactical Neutralization is common to those familiar with Far Eastern philosophical disciplines and directions in the martial arts. Traditionally representative of the endeavor to achieve harmonization of mind, spirit, and body it has come to dignify more contemporary applications and hopefully, the institutionalized aspirations of law enforcement.

We perceive the symbol as indicative of the attitudes, perspectives, and orientations of which the Tactical Neutralization Techniques paradigm consists. This is intrinsically expressed in the nomenclature "tactical neutralization," which is manifested in the illustrated hand juxtaposition. Although the symbol was approached with desired and designed ambiguity; force yet discretion; power yet restraint; control yet tempered compassion; and aggression potential, yet the power of passivity, are but a few representations of the conceptual meaning of our selected symbol.

The system does include four levels of performance proficiency represented by color status designation, illustrated as follows:

LEVEL 1

Blue—Initial certification of proficiency in Tactical Neutralization Techniques under the direction and review by authorized instructors.

LEVEL II

Brown—Superior proficiency in Tactical Neutralization Techniques designated under the direction and review by authorized instructors.

LEVEL III

Black—Continued superior proficiency of Tactical Neutralization Techniques in the field of enforcement. Black status will be awarded based upon an individualized evaluation format under the direction and review of authorized instructors, provided the Individual has attained at least one year of field experience and has made a structural contribution to the Tactical Neutralization Program.

LEVEL IV

Red—Superior proficiency within all designated areas of Tactical Neutralization and the proper credentials to instruct others in these skill activities. Evaluation of applicants will be made by authorized instructors who have attained red status, one of which must have sponsored the applicant.

In conclusion it should be noted that in its totality Tactical Neutralization encompasses the entire continuum of enforcement encounters and the varieties of subject actions. This continuum founds itself upon the officer's multiplex roles of encounter so vital, and yet so varied, in the enforcement environment. This continuum stretches from acknowledged acceptance of an officer's actual or even perceived presence, up to the specific scope of that officer's exercise of his remotest responsibility . . . that of deadly force. A separate chapter will follow, describing the components embodied in this control continuum, as well as its action applications.

This continuum design should be ever present in the mind of the officer and the reader and the inter-relationship of previously distinct studies of social encounter, crisis intervention, firearms, legal issues, control tactics, etc., should fuse together for even greater understanding and utilization of Tactical Neutralization Techniques.

Professor Gregory J. Connor

Professor Connor has been a police officer and police trainer for over 20 years. He is the co-developer of the Tactical Neutralization Techniques program at the University of Illinois Police Training Institute. Greg Connor is recognized as an expert nationally and internationally in the areas of police use of force and control tactics. He holds a Black Belt in Isshinryu Karate and is a Licensed Instructor with the United States Karate Association. Professor Connor has published numerous articles on a variety of police issues and texts on the Police Yawara, PR-24 Police Baton, and Women's Defensive Tactics. Mr. Connor has a B.S. degree from Michigan State University, a M.A. degree from the University of Illinois, and is engaged in advanced graduate studies at the University of Illinois.

Matthew D. Summers

Matthew D. Summers has been a police officer and police trainer for the past eight years. He was the co-developer of the Tactical Neutralization Techniques program while an Instructor at the University of Illinois Police Training Institute. Matt Summers has an extensive background in a variety of martial arts, holding a Black Belt in Tae Kwon Do. He has worked at the Municipal, State, and Federal levels of law enforcement, as well as in private security. Mr. Summers holds a B.S. and M.A. degree from the University of Illinois and is continuing advanced graduate studies.

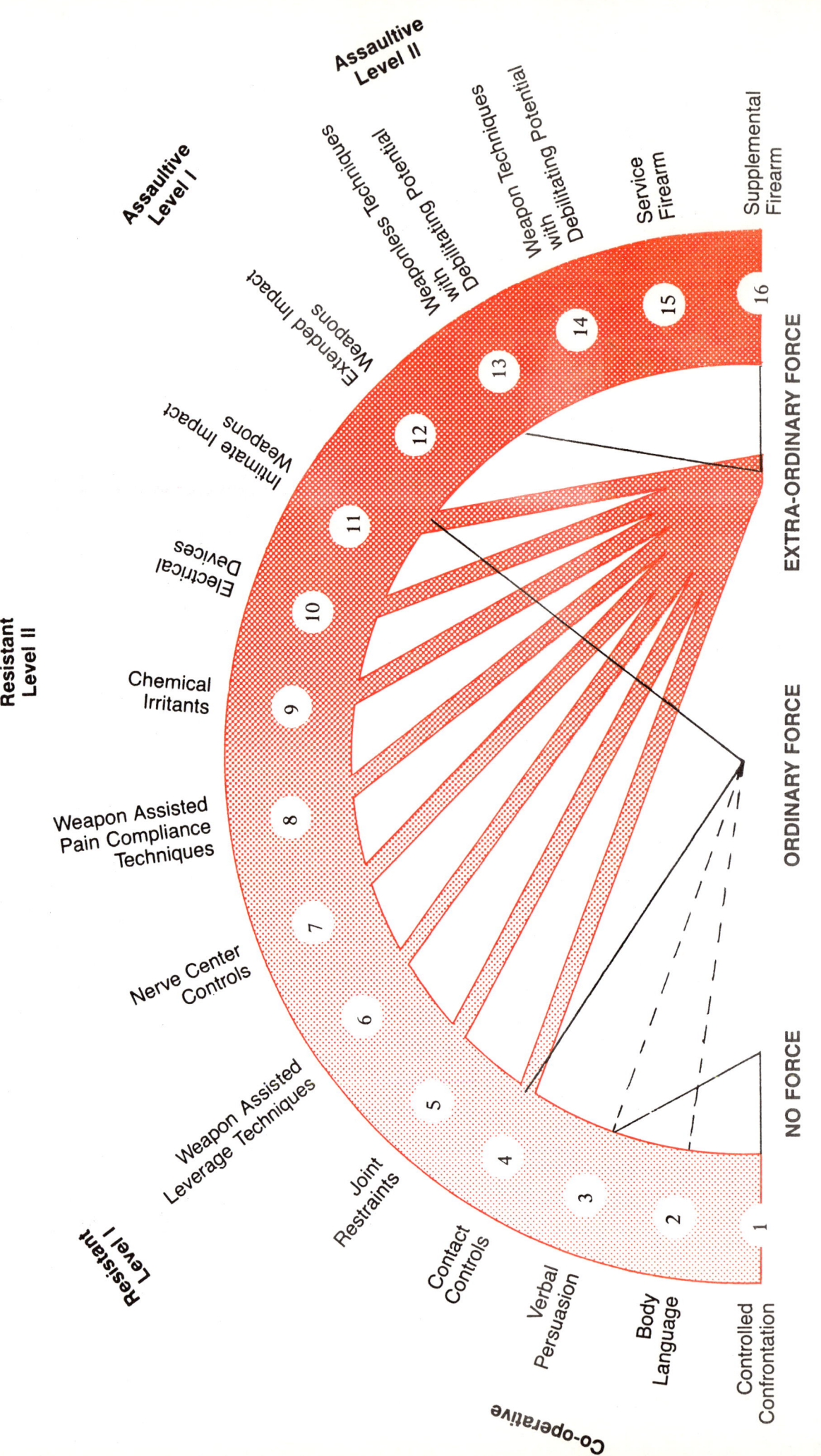

Introduction

USE OF FORCE: CONFRONTATION/CONTROL CONTINUUM

    A critical issue area confronting contemporary law enforcement centers around the use of force. In most states specific criminal statutes spell out an officer's rights and responsibilities relative to force utilization. However application of these statutes in action terms still seems difficult to comprehend by many officers, and likewise, at times too complex for departmental comprehension in the implementation of effective policy.

    A variety of instructional designs have been created to enhance understanding and application approaches. As a result of these efforts and additional legal and learning research, the Use of Force: Confrontation/Control was developed. Although other use of force graphics are available, this continuum does possess a unique simplicity as well as a multiplicity of functions.

    The Use of Force: Confrontation/Control Continuum graphic requires an explanation of its features. Beginning on the left side of the continuum, faint red coloration depicts the use of no physical force within a confrontation setting. As is the case in the vast majority of citizen encounters physical force is never used by the officer. Force in these day-to-day situations is neither reasonable or necessary and is usually negated by a professional officer with a controlled demeanor. The trained officer can initiate manipulative behavior with body language and/or verbal persuasion.

    This faint red coloration of no force sharply ends once physical rather than cognitive or verbal manipulation occurs. The commencement of ordinary force, or that physical force not likely or intended to cause death or great bodily harm, is indicated with a light red color eventually increasing in its red hue as ordinary force options increase in degree or physical coercion.

    Finally, the continuum extends into the far parameters of physical force likely or intended to cause great bodily harm. This level of force is displayed with a continuing red coloration increasing to its eventual full color intensity.

    Other features that assist in this graphic illustration of force are three types of lines. The solid lines clearly designate either no force, ordinary force, or extraordinary force. The wider bar-like lines are drawn to those weapon/tactic areas that have proper application in the realm of ordinary force but if abused could become extraordinary force.

    For instance, a joint lock technique is normally used to secure a resisting subject and diffuse aggressiveness well within the parameters of ordinary force. However, if the officer decides to intentionally apply pressure on the joint in order to dislocate it, such action would fall into extraordinary force.

    The dash lines designate the illegal extension of a non-physical force alternative from its normal positioning into a criminal stature via excessive or abusive conduct by the officer. In such case an officer's verbal comments to a citizen could be intimidating, or if his actions were regarded as threatening in nature, they could actually be assaultive. Within the arc are 16 numerals which identify either a weapon or technique option that is generally available to most law enforcement agencies as well as a professional consensus as to their approximate force location on the continuum arc.

    These headings are generalized in their positioning a profile, and are to be utilized for illustration rather than legal, procedural, or tactical illumination.

    1. Controlled Confrontation—represents the vast majority of police-citizen encounters. Other than the authority inherent in the position as perceived by the citizen, the presence of power, symbolic influence of uniform, the squad car, etc., actual physical force remains non-existent.

    2. Body Language—as much as 70% of communication occurs through non-verbal channels. A reluctant individual may be convinced into compliance by the actions of the officer; a body shift toward the offender, intensified eye contact with the individual, etc.

    3. Verbal Persuasion—a noncompliant individual can be verbally convinced by the officer into a pattern of control. The direction of the structured conversation may be toward the futility of resistance, the reassurance of the officer's authority, or utilizing time for more assistance to arrive, etc.

    4. Contact Controls—a reluctant individual can be brought into compliance with mere placement of "hands on" and the assumption of an arrest posture. Generally the officer assumes control of the subject's arm while positioning himself to the rear/side of the subject. This contact allows potential flexibility of resistance if continued and in many cases adds a degree of futility to further subject aggression.

5. Joint Restraints—a mildly resistant subject can be placed into a controlled posture with a leverage domination technique. Usually the wrist or elbow joint are the focus of a variety of restraint tactics which can grow in tension or pain compliance reciprocal to subsequent resistance.

6. Weapon Assisted Leverage Techniques—a variety of weapons can have a primary or secondary capability of assisting the officer with a mechanical improvement in control efforts.

7. Nerve Center Controls—resistant subjects may necessitate additional areas of pain compliance cncentrated upon isolated nerve centers throughout the body until an acceptable threshold of pain is experienced as a simultaneous control tactic is employed.

8. Weapon Assisted Pain Compliance Techniques—to facilitate more rapid and/or effective pain compliance against a belligerent subject, weapons; including yawara sticks, batons, etc., can be utilized to conduct controlled strikes or pressure stimulation into traditionally recognized nerve enters, muscle groups, or joint locations in order to facilitate control.

9. Chemical Irritants—individual aerosol canisters containing a tear gas formulation can be used to subdue an aggressive, noncompliant subject. They become more ideal for the subject who is increasingly violent in his resistance while providing less opportunity for injury to either party.

10. Electrical Devices—a type of mechanism very effective on physically agitated individuals are the elctrical based weapons. Via high voltage and safe, low amperage, these devices override the body's nervous system and cause muscle dysfunction allowing for directed submission until control is assumed.

11. Intimate Impact Weapons—with a violent subject, striking or punching techniques can normally be directed toward nerve centers and are made more efficient with the use of small weapons under the generic heading of yawaras or yawara sticks. The tactical use of these weapons necessitate close contact to be functional but control and power tactics are significantly improved with a focused targeting of impact.

12. Extended Impact Weapons—an extremely violent subject endangers close proximity techniques therefore strikes or thrusts can be best made by extension designed weapons under the generic heading of batons or nightsticks. With distancing, the officer has the ability to control the subject's violent actions with a minimized risk potential.

13. Weaponless Techniques with Debilitating Potential—extremely violent subjects must be immediately subdued to minimize risks. These techniques attack life essential centers of the body and therefore, allow for total control. Perhaps the most widely used technique is the lateral vascular neck restraint. Once initiated it can result in unconsciousness within several seconds, or possible death if tension is continued.

14. Weapon Techniques with Debilitating Potential—again if distance is essential, a baton power struck to the knee joint can cause joint dislocation and prevent further movement.

15. Service Firearm—when necessity requires immediate curtailment of deadly, aggressive behavior the use of the service weapon in combination with the center-mass targetng of the perpetrator may be the most acceptable alternative.

16. Supplemental Firearms—a violent situation involving life threatening risk to the officer or others may necessitate the use of firearms with increased power or flexibility. Normally these incidents relate to situations involving hostage encounters, barricaded subjects, etc.

Across the top of the arc, five categories of subject action have been provided and defined for ease in understanding and application of reciprocal force. It should be noted that definitions and/or demarcations of such human conduct are in reality difficult if not impossible to profile with great specificity. However, broad categories of action on the part of the individual and an appropriate reaction on the part of the officer can be of assistance in transitional training for the new or veteran officer in this critical area.

Cooperative

In the course of normal patrol, the officer's contact with the general population is vastly more positive than negative, and likewise, much more prone to nonforce than forceful confrontation.

Statistically our feeling is that an officer should see an enforcement range of approximately 95-97% in the realm of cooperation and control.

Even in the exceptional cases, at times the officer may simply re-adjust his spatial positioning or elicit greater eye contact (body language methods) and gain the reluctant compliance of the individual. Or the non-compliant individual may have the officer's request repeated or verbally convinced that increased reluctance need not progress to resistance (verbal persuasion methods) culminating in eventual compliance.

Resistant (Level I)

In those exceptional confrontations, individuals may offer the preliminary cues of resistance. Their opposition to the lawful request for compliance by the officer may be exhibited in a variety of manners.

Verbally, the individual may repulse the officer's efforts. His abstinence may be purely inactive and he may go limp, negating the officer's direction.

This level of inert resistance, however, is still legally within the parameters of resisting, and during an arrest confrontation, selected control options could include 2-7.

Resistant (Level II)

Resistance in this classification now becomes active in its scope and intensitty. The indifference now has become expressed via physical defiance. The subject may turn away from the officer and attempt to leave the scene. He may pull out of the officer's attempts at contact controls. He may attempt to remain stationary by grasping onto the steering wheel, or he may attempt to escape by placing barriers prohibiting the officer's pursuit. Critical to the categorization of Resistant (Level II) is that no direct force or violence has been directed toward the officer. It should be noted, however, that pulling out of a control technique could directly or indirectly injure the officer and, therefore, subsequent techniques deployed attempting to gain control of this subject could legitimately potentiate into a higher level of noncompliance.

Generally, control options 2-10 are legally within the parameters of selection in this classification of an active resistant subject.

Assaultive (Level I)

In this level the officer's attempt for lawful compliance has met with active, hostile resistance expressing itself in the form of physical attack upon the officer.

The criteria for the severity of the battery to the officer is its reasonable potential of its action resulting in harm to the body. For instance, within the confines of a family dispute, an officer would be hard pressed to call a woman an Assaultive (Level I) subject for her throwing a pillow at him from across the room. However, if that same woman were to throw a frying pan at him, it would be reasonable to assume that some bodily harm could befall the officer.

The resistance keys upon the direction of the violence as well as its intensity and, therefore, includes a large realm of resistance activities. In each case, the specific judgment as to overall scope of the violence must err on the side of officer safety. Control options 2-12 are available to the officer.

Assaultive (Level II)

In this category exists the least frequently encountered individual, but obviously the subject of most concern. Here lies the violent subject who by his actions creates a reasonable assessment that his non-compliant activity has the potential to cause the officer great bodily harm and even death.

Although such activity is most severe in its impact, its assessment many times appears to be the simplest. Most individuals can readily, almost instinctively, perceive when their actions are in defense of life and, therefore, justified in force options 2-16.

It should be noted that in all categories, the officer's training perception and appropriate anticipation must be structured into our subjective scale of reasonableness. These traits will be enhanced as via your street and tactical experience.

This training methodology applied to the use of force has been designated as the progressive application of force. In action terms of this process is founded upon a tactical perception on the part of the officer in reference to the clues and cues of subject behavior and the appropriateness of the officer's reaction to these traits. Our goal is embodied in a portion of our model continuum graphic, the realization of control.

Although the theory of progression application of force appears to focus upon a format of action/reaction, an important element of pro-action is the essence of this approach. The pro-action element is fostered by an increased awareness by the officer of the subject's behavior and the assessment of relevant risks to the officer.

The skills of street survival blend the cognitive as well as the coordinated, the tactical with the tacit, and the courageous with the compassionate. Control is an ever changing facet of human interaction. Our actions must be as defined and dynamic as our goal for societal safety.

Chapter 1

WHY PHYSICAL TRAINING AND PERSONAL DEFENSE

Preparatory education and ongoing training are no doubt the most essential elements as you begin or continue your career in law enforcement. Our program design includes a tricotomy of instructional criteria composed of text materials, exercise, and practical tactical simulations. You, the student officer, must comprehend the presented material, regiment your day to include these basic exercises, and most importantly, practice and master the police tactics into a pattern of professional presence.

Let us begin our information exchange by giving a comprehensive answer to our lesson's title question, "Why physical training and personal defense?"

Job Assignment

Job assignment can be addressed on two fronts: personal defense and public protection. Police work by its nature has a potential for physical confrontations. Included under these physical confrontations most experienced officers can recall violent citizens, fleeing youths departing from their stolen auto, application of CPR upon a dying citizen, or the race against time attempting to pry a seriously injured child from the mangled debris of a traffic crash. We would be romanticizing if we pretended these events were happening with daily regularity, but we would be remiss in failing to prepare you with the capacity to physically meet and manage each of these incidents.

Fortunately, our knowledge of the types of incidents is no longer without research basis. Nearly a year of study by the Los Angeles County Sheriff's Department, has established twelve incident-related skills in rank order. Although variance certainly occurs between small, medium and large agencies as well as duty assignments, this listing is still valuable to you as a recruit in structuring the specificity of your physical exercise program. The following incident-related skills include:

• Two hundred sixty-seven climbing incidents were reported. The most common obstacle climbed was a six-foot concrete wall or a ten-foot chain-link fence. Most climbing was necessary during the investigation of a crime.

• Two hundred thirteen running incidents were reported containing from 0 to 2 turns. The distance most often run was between thirty and eighty feet. Most running was necessary for the apprehension or control of a suspect.

• One hundred forty-nine jumping incidents took place. The distance most often jumped was between three and seven feet; the direction was down and was necessary for the investigation of a crime.

• Lifting was next with 124 responses. The most common weight was 175 pounds or more, and the most common object lifted was a person.

• Balancing incidents amounted to 84 responses with a distance between nine and twenty-seven feet. Tthe most common object balanced on was a concrete wall and was necessary during the investigation of a crime.

• Pulling accounted for 73 responses over a distance between three and fifteen feet. The most common weight was 175 pounds or more. The most common object pulled was a person for their apprehension or control.

• Pushing, with 66 responses, was most often done on a car over a distance of three to fifteen feet.

• Carrying received 51 responses. The object carried was undifferentiated over a distance of nine to twenty-one feet. The most common weight was 25 to 75 pounds.

• Wrestling received 50 responses. The most common opponent was an adult male between 150 and 175 pounds.

• Crawling was next with 27 responses, over a distance of three to nine feet through some type of obstacle and was necessary for the investigation of a crime.

• Dragging with 24 responses consisted most usually of dragging a person weighing 175 pounds or more for a distance of nine to twenty-one feet for purposes of apprehension or control.

• Hitting or kicking was last with 22 responses. Hitting of a person for their apprehension or control was the most frequently cited response.[1]

On the other hand, we have selected our career, we pinned on the badge and we, officer by officer, have created an aura of action toward the public. Our underlying task is control; control of a society that community by community outnumbers us by often more than 1,200 to 1. Although in the street confrontations of the 1960's, the public discovered that we were not omnipresent, and more injurious to our image, not omnipotent, they still have lingering expectations of heroes in blue. At times, they expect you to accomplish super human feats. You still comprise the thin blue line that separates the peaceful and free-living citizenry from the ravages of the ever-growing criminal faction. Public expectations should be looked upon as not just a challenge, but additionally as an opportunity to upgrade and enhance your representation in society.

Personal Enhancement

There is no better time to consider your career potential than now as you enter police work. Experienced officers agree in principle with leading educators when they voice the need to develop the totality of self. An easy illustration of the old axiom, "sound body equals sound mind," is generally the productivity of the recruit officer during his weeks at the Institute. The mind is initially motivated, perhaps out of fear of job loss, career sincerity, etc.; however, as the days and weeks progress, the average officer spends less and less time and energy on classroom activity, his attention span wanes, and his supportive study away from the classroom dwindles. This mental energy decline is paralleled in physical deterioration, made even worse by a probable weight gain.

"Physical fitness is not only one of the most important keys to a healthy body, it is the basis of dynamic and creative intellectual acttivity. The relationship between the soundness of the body and the activities of mind is subtle and complex. Much is not yet understood. But we do know what the Greeks knew: That intelligence and skill can only functiton at the peak of their capacity when the body is healthy and strong; that hardy spirits and tough minds usually inhabit sound bodies."[2]

In actuality, the self-image, body image, and self-estimate are all cross related and correlated to movement depravation and physical deterioration. If an individual is ranked or ranks himself lower in any of these assessment areas, he will probably have a poor self-concept. An individual who has a healthy body image, self-image, and self-estimate is more likely to attempt new physical activities, and participation in these can provide opportunities for social recognition and development of friendships.[3]

Simply look around your department and see what happens time after time, officer after officer; first it's expressed through pseudo-fatigue, then a lack of motivation eventually ending with a decline in morale. To us, boredom results in many more deaths (terminal to career potential) than the bullets we fear . . . and disproportionately fear so much!

Personal/Professional Status In The Community

Here the key word is status, or better yet, your image, as a police officer. It is no surprise that one can look professional and gain reciprocal compliance. It has been shown that, "A person who is in poor physical condition becomes fatigued more easily than a person who is in good physical condition. A fatigued person, in making judgments in stressful situations, will rely more on his personal instincts and prejudices and proportionately less on the training he has received. It is vital that police officers at all times exercise judgment in accordance with Department policy."[4] If we are to police a democratic society, we must control the minds of at least the majority through respect, rather than fear. "The low self-esteem person defends himself by hostility, withdrawal, excessive assertiveness in the use of power, insults to others, etc. As the threat increases his anxiety, his thinking may become more rigidified and his solution of the problems at hand become less effective."[5]

"The continual striving for a proof of virility provides a need to demonstrate one's masculine superiority through physical altercations, capacity for drinking and sexual prowess, which in turn, can lead to considerable embarrassing acting out or be a source of behavior that leads to aggravation of family problems, citizen complaints, a greater readiness to shoot, generally a more aggressive, assertive, competitive approach leading to confrontations and to further physical and emotional injuries and destruction of equipment and uniforms."[6] In our opinion, that control can best be accomplished via imitation rather than innovation. The juvenile seeking a model should need only to look at the most visible and viable portrayal of government in his community . . . the officer in uniform. The positive effect of the uniform can best be maximized by a well-proportioned, well-conditioned body.

Valid Mechanism For Stress Release

If the physiological balance of the body is disturbed, the individual experiences stress. In simplest terms, stress is excess energy created in the body, principally by fear. These fears can have a psychological as well as a physiological basis but regardless, the body does not differentiate the source of the stressors. Therefore, the body's balance is identically related in reaction until a balance is resumed.

Hans Selye, a Canadian physician and endocrinologist, who is thought of by many as the "father" of contemporary stress research, has depicted the body's stress reaction in three stages. In stage one, the *alarm* stage, the body reacts to the stressor. Here, the body's sympathetic nervous system and the adrenal glands are stimulated into instant action. The following body changes occur:

1. The heart rate and stroke volume are increased, which results in an increased cardiac output.
2. Blood vessels in the skin, kidneys, and most internal organs become constricted, which decreases the blood flow to these areas; blood vessels in the skeletal muscles become dilated, which increases the flow to them.
3. Systolic blood pressure and the volume of blood circulating per minute are increased.
4. Secretions and *peristalsis* of the digestive glands are decreased, and thus there is decreased digestion.
5. Liver glycogenolysis (breakdown of glycogen to glucose) is increased, which results in more blood glucose.
6. The breakdown of adipose-tissue triglyceride is increased.
7. The rate of ventilation is increased.
8. Muscle tension is increased.[7]

Soon after the stress or stressors have alarmed the physiological networks, a second stage of *resistance* is initiated. Short or long term body adaptation occurs, depending on the stressor.

However, stage three, that of *exhaustion*, takes place if no adaptation takes place or the adaptive mode viable during the resistance stage eventually becomes exhausted.

The process of repeated adaptive reactions or the maladaptation can throw the officer into major categories of stress diseases and disorders. Stress is the reason, named time and again, by researchers of occupational hazards as a major dibilitating factor in the police officer's job.[8]

"The body's faulty adaptive reactions to stress appears to encourage various maladies, including emotional disturbances, headaches, insomnia, sinus attacks, high blood pressure, gastric and duodenal ulcers, rheumatic or allergic reactions and cardiovascular and kidney disease."[9]

"Stress is a problem that has been honed to a fine point in America. There are two kinds of stress, acute and chronic, and they can be either physical, emotional, or both. Acute is the more dangerous of the two. It could mean sudden death. It's a fight-or-flight moment, and your body—specifically the sympathetic nervous system—tries to accommodate the emotional emergency with a fresh supply of hormones to stimulate the heart to a higher work load. If your heart is not in condition to take on this higher work load and is forced to pump blood at a faster rate, if the situation continues long enough, or the rate goes high enough, it could give up. Heart attacts are not uncommon at moments like this. It's called a "sympathetic storm." The sympathetic system produces a rash of hormones, more than the parasympathetic system can neutralize, and more than the heart can safely handle. These storms, however, do not have to be entirely emotional. Acute physical stress can produce the same result. An obvious example might be a holdup. Suppose it happens to you, you start running, and the gunman starts shooting. The emotions of the moment and the physical stress of running might be enough to drop you quicker than one of the bullets."[10]

Many of the stresses an officer faces in his environment become eventual distress that have the potential to afflict both physical and emotional health.

Psychological stress produces not only what is commonly thought of as being mental and emotional disturbances, neurosis and psychosis, regressions, brain damage—related problems known as organic brain syndromes, and so-called traumatic neurosis also known as combat neurosis, gross stress reaction, or transient situational disturbances often resulting from life and limb threatening situations or other line-of-duty crisis, but also produces a whole gamut of psychophysiological disturbances that, if intense and chronic enough, can lead to demonstrable organic disease of varying severity. A list of such psychophysiological conditions that lead to medical disorders of the skin such as neurodermatitis and atopic dermatitis; of the musculoskeletal system such as backaches (the low back syndrome), muscle cramps, tension headaches, stiff neck; psychophysiological respiratory disorders such as bronchialasthma, hyperventilation syndrome; psychophysiological cardiovascular disorders such as high blood pressure, tachycardia, gastrointestinal disorders such as peptic ulcers, chronic gastritis, ulcerative and mucous colitis, constipation, hyperacidity, pyloric spasm, heart burn, irritable colon, gastroesophageal reflex; psychophysiological genitourinary disorders such as physiological endocrine disorders such as diabettes mellitis, thyroid disorders, adrenal disorders, pituitary disorders, menstrual disorders, and other sexual hormone disorders. There is also increasing evidence that the occurance of industrial accidents themselves are often stress-related; this has been called the "accident prone. . . ."[11]

As an officer gets older, the need for a preventive health program becomes more evident.[12]

However, it can be argued, "that today's technology, rather than decreasing, has increased the need for officers to keep constant vigilance over their biological systems and maintain themselves at highly functional levels."[13]

"In the words of Public Health Services' Forward Plan for 1977-1981: Habitual inactivity is thought to contribute to hypertension, chronic fatigue and resulting physical inefficiency, premature aging, the poor masculature and lack of flexibility which are the major cause of lower back pain and injury, mental tension, coronary heart disease and obesity."[14]

Moreover, police officers are among the highest in incidence to America's number one killer, heart disease, according to the United States Public Health Service. Constant coronary risk factors identified include high blood fats (i.e., cholesterol and triglycerides), high blood pressure (hypertension); cigarette smoking; obesity; heredity; stress (due to role overload, pressure from time schedules and deadlines, etc.); physical inactivity; other (i.e., sex, age, personality, race and subculture).

The American Heart Association claims that well over 40% of all deaths in men between the ages of 40 and 59 are the result of coronary heart disease. In an even wider scope, 52% of all deaths in the US in 1977 were due to some form of cardiovascular disease. Researcher E. D. Michael reported in the *Research Quarterly* that moderate exercise actually provides a mild motion of stress for the adrenal glands, which helps to condition and fortify them so they can more effectively reach a proper adaptive mode for more severe stressors.[15]

Studies have repeatedly indicated that fat is cleared more rapidly from the blood after exercise and that a program of physical training can preserve this effect. This clearing capability can serve to prevent arteriosclerosis (an occluding process occurring within the blood vessels) which is the major contributor to the development of coronary heart disease.

Physical exercise is one of the best methods of adaptation an officer can select to keep daily stress from reaching harmful levels of distress. Exercise programs are not only a positive adaptive avenue for the physiological threats of stress; numerous studies reveal that moderate exercise, sports, even walking or jogging have great value in the release off mental tension. Habitual worry or anxiety is temporarily forgotten, and physical tensions arising from these concerns are released.[16]

Footnotes for Chapter 1

[1]Osborn, Gary D. "Validating Physical Agility Tests," *Police Chief,* January, 1976, pg. 44.
[2]Kennedy, J. F. "The Soft American," *Sports Illustrated,* Dec. 26, 1960, pg. 159.
[3]Miller, David and Allen, T. Earl. *Fitness: A Lifetime Commitment,* Minneapolis, Burgess, 1979, pg. 3.
[4]"Physical Fitness," Los Angeles Police Department Training Division Bulletin, Vol. 3.72 F/4, pg. 5.
[5]Stotland, E. "Self-Esteem and Stress in Police Work," *Job Stress and the Police Officer,* Washington, D.C., U.S. Government Printing Office, December, 1975, pg. 3.
[6]Jacobi, J. H. "Reducing Police Stress: A Psychiatrist's Point of View," *Job Stress and the Police Officer,* Washington, D.C., U.S. Government Printing Office, December, 1975, pg. 91.

[7]Anthony, C. P. and Kolthoff, N. J. *Anatomy and Physiology,* 9th Ed., St. Louis, C. U. Mosby, 1975.

[8]Goodin, C. "Opening Remarks," *Job Stress and the Police Officer,* Washington, D.C., U.S. Government Printing Office, December, 1975, pg. 17.

[9]Reiser, M. "Stress, Distress and Adaptation in Police Work," *Job Stress and the Police Officer,* Washington, D.C., U.S. Government Printing Office, December, 1975, pg. 17.

[10]Cooper, K. H. *Aerobics,* New York, M. Evans and Company, 1968, pg. 122.

[11]Jacobi, J. H. Op. Cit., pg. 86.

[12]Pollock, M. L. "Physical Fitness Training Programs," *The Police Yearbook,* Gaithersburg, Maryland, IACP, 1977, pg. 266.

[13]Woods, M. D. "Physical Efficiency and Tension Control: Two Factors in Police Officer Safety," 1975, National Safety Congress Transactions, Vol. 8, Construction, Public Employee, Chicago, National Safety Council, 1976, pg. 85.

[14]Keeler, R. O. "Address to 1976 Blue Shield Annual Program Conference, Chicago, October, 1976, pg. 2-3.

[15]Michael, E. D. Stress Adaptation Through Exercise, *Research Quarterly 28:* 50-54, 1957.

[16]Byrd, O. Studies on the Psychological Value of Life—Time Sports, *Journal of Health, Physical Education, Recreation, 38:* 35-36, Nov.-Dec., 1967.

Chapter 2

STATUS OF "FUNCTIONAL FITNESS"

Contemporary interest in physical fitness among the general population has never been more diversified and more intense. For the police officer, this interest is closely aligned with occupational necessity.

Fitness may mean the difference between control of a situation or a violent overreaction, between a positive encounter or a negative confrontation, and perhaps, ultimately, between life or death.

Physical fitness is perhaps best defined as a status of positive health and peak body efficiency. "Functional fitness" is seen as the ability for today's police officer to have the energy, expressed via motivation and action, to accomplish, the tedious activities of policing; to guarantee a reserve capacity for those uniquely police emergencies that can happen anytime and anywhere; and at times forgotten, but critical to anyone in the stress-filled occupation of law enforcement, the interest and ability to become committed to family leisure activities.

The foundation of most, if not all, organized fitness programs is exercise. Exercise has been called the major therapy for the ill and the master conditioner for the healthy. The human body has been compared to a machine in this regard with one unique variance. The difference is that unlike the machine that wears out after use, the human body thrives upon and actually improves with its use. Therefore, inactivity, for whatever the cause, can result in body deterioration. It has been said that physical deterioration of the body is the worst disease in the United States today. It is so prevalent that if you are over 25 years of age, there is a 50% or more chance you are suffering from some form of the disease.[1]

For your own interest in the self-assessment vein, here are some of Dr. Allman's "red flags," or indicators, of this disease of physical deterioration:

1. *Accumulation of noticeable fat, especially in the lower abdominal area.* Here at the Police Training Institute, a study was conducted in cooperation with the University of Illinois College of Physical Education on police officers. The results showed, "the law enforcement group was below average in this ratio (strength divided by body weight) that is, they were excessively fatty."[2] Our continuation of this research effort via a visual assessment would minimally depict at least one-third of the officers in the excess adipose tissue category.
2. *A loss of muscular strength.* Circumstances may occur on the street which require you to exert force. Whatever your natural muscular ability, the capability of calling upon its peak potential may be your only alternative.
3. A loss of endurance. Policing requires a capacity for continued exertion for prolonged periods. With growing regularity, officers are called upon to recover or sustain life using cardio-pulmonary resuscitation techniques. Your endurance is tested within minutes after commencing the activity. If you are unfit, your duration capability for life support is sharply curtailed.
4. *Slowed reaction time.* Challenges to your reaction time can take the form of a left jab, roundhouse right, a kick to the groin, a car door opened rapidly, a motor vehicle darting in front of you in its entry into traffic, or failure tto detect that body movement in your building search. The retardation of reaction time will cause you to, at least, have less opportunity to "think" out your reaction or, at worst, become a victim of what could have been a controllable activity.
5. *Loss of flexibility.* Flexibility is the ability of the body to stretch joint regions to provide for broad, efficient movement. While on patrol, the restrictions upon body movement are most apparent when applied to the officer's lower back due to his seated position for hours. No wonder lower back pain is the complaint of the 35-year old police officer and the curse of the 20-year veteran.

At this time, the most obvious question is "Why? . . ." Why would people who are engaging in an occupation like law enforcement neglect their own fitness levels? In our opinion (over a period of the last 20 years), if anything can be noticed about traits in regard to recruit officers, one of the most obvious observations would be the deterioration of fitness levels.

But we are predictable in nature; rather than admitting the truth as confirmed by a glance and initiating our own program of self-conditioning, we instead erupt with excuses. The most frequent excuses are the following:

Lack of facilities. Like most members of today's structure-oriented society, officers express an aversion to exercise if no gymnasium is available. Many complain that their departments fail to provide an exercise room or a pool, etc. Perhaps we are old fashioned, but have entered the realm of law enforcement presuming it to be a position of privilege.

In order to stay in shape, some fellow officers sought out facilities because they saw physical conditioning as being a *significant* factor toward safety. However, as the assault rates upon officers continue to rise, many others continue to rationalize.

Next on the priority list of excuses is the voiced *lack of equipment.* If functional fitness is the end, what activity is considered the best overall means? Most sources agree it is running . . . from jogging to racing. The equipment necessary includes merely a quality pair of running shoes at a cost of under $40. Wholesale merchandising has brought sports equipment, once financially removed, more available to our incomes.

The next excuse, one which seems to loom well on into an officer's career, is *lack of time.* No doubt, our career choice has placed us into an environment fraught with fluctuating schedules that can conflict with a program of self-conditioning. However, most experts agree that the time necessary to maintain a level of functional fitness is less than an hour. The time is there, just make the effort to find it.

A *lack of fundamental physiological knowledge* has limited initial interest in conditioning for some, and inhibited the continuation of exercise activities for others. The lack of knowledge includes, but is not limited to, ignorance of muscle development, muscle fatigue, proper nutrition, exercise techniques, and the modeling of an exercise format for the individual.

Last of the rationalization factors, but by far not the least, is the *pressure,* or in some cases *lack of pressure from peers.* Even in recruit school, at times, officers who engage in routine exercise activities are subjected to ridicule by nonparticipatory types. This jousting or jealousy may not subside once he goes back to the department. The motivation must be self-inspired. Long term motivation is essential to achieve and maintain "functional fitness." This motivation in the final sense must be self-inspired.

Footnotes for Chapter 2

[1]Allman, Fred L., MD. Executive Fitness Desk Diary, M.B. Productions, Company, Dallas, Texas, 1969, pg. 7.
[2]Pohndorf, Richard and Cathey, Richard. "Fitness Changes During a 14-Week Basic Law Enforcement Training Program," FBI Law Enforcement Bulletin, federal Bureau of Investigation, United States Department of Justice, Washington, D.C., January, 1975.

Chapter 3

PREPARING TO EXERCISE

If you are not currently in a structured fitness program, no better time exists than right now. Why not begin a formal exercise effort modeled after those of other police agencies? Perhaps the next several weeks of regimentation can act as an increased incentive toward an avenue of fitness.

Major categories worthy of discussion in your preparation program include:

Motivation—be it pride, the pursuit for personal improvement, whatever the basis; as long as the incentive is durable enough to last through the sweat, muscle fatigue, and the boredom that eventually attaches itself to exercise.

Goal—Make it reasonable, don't attempt too much—too soon. Seek the practical goal of optimum fitness, maximizing your body's resources.

Medical evaluation—before any actual activity commences, it is highly imperative to receive a medical evaluation so that a competent physician can safely set the initial parameters of exercise.

Time—here the most important consideration is consistency, finding the right time of the day for you in your pattern of life. Exercise for some may be the eyeopening event of an early morning run, for others it may be the alternative for a lunch in our battle of the bulge, and for still others it is the manner in which we relieve the stresses built up over the day. Keep in mind that in order to secure that essential trait of consistency, others, individuals and institutions, must be considered. Individuals in your life pattern—wife, children, neighbors—are critical in your selection of a time to work out. The institution to consider, of course, is the department since it can wreck even the most rational of plans. On those occasions of shift changes, overtime assignments, or other police matters too numerous to mention, you may have to skip a day or become overly fatigued. Push yourself to resume the exercise sequence as immediately as possible since we all have an increased ability to avoid work and/or rationalize our way out of a habit of exercise.

One last note as to the time when specifically one should not exercise. That time is immediately after a major meal for at least two reasons. Most important is that digestion requires increased demands on the heart; if you exercise during the time of blood demands on the heart, it is unnecessarily overtaxed. Also, if strenuous activity follows a major meal within less than a 1½ to 2 hour period, serious indigestion is not uncommon.

Place—the only requirement for an exercise environment is that it is conducive to physical activity. Distractions to your cncentration upon a fitness program can include inappropriate temperatures, restrictions on space, or a limitatin to a necessary degree of privacy. Take the time to select and outfit your workout area and in doing so, assure yourself of long-term participation in an exercise program.

Frequency and Duration—most experts state that those who are beginning to exercise should conform to a progression format of exercise. It is suggested that the efforts of exercise should be extended to at least five times per week (not just days off, since leave days may feature an above average of caloric intake.)

The duration of the daily training session generally meets the suggested length of one hour. This hour is divided into four segments including periods of acceleration, training activity, deceleration, and revitalization.

Acceleration—is simply the warm-up activity which allows the person to minimize muscle and joint problems. These stretching exercises, activity to stimulate heart and lung activity, can be successfully accomplished in approximately 10 minutes.

The training activity defines the major portion of the workout, be it running, swimming, a game of racquet ball, etc. To conform with our one hour of exercise activity, this period should extend for 20 minutes, with 8-10 minutes inclusive where you should push yourself with levels of tolerance so as to increase your endurance.

The period of deceleration is just as critical for your well-being as the warm-up and for the same duration of 10 minutes. Stretching exercises in conjunction with walking will gradually return the body to its healthy metabolic state.

Revitalization—commences once your body has cooled down via a warm, then gradually cooling shower, for a period of 15-20 minutes. This warm water immersion aids circulation and thus accelerates the body's recuperative efforts.

Symptoms to stop exercise—finally for your safety you should constantly perceive any signals your body will give if the exercise is beyond your body's tolerance levels; these signals include light-headedness, loss of muscle control, chest pains, loss of breath and nausea.

Chapter 4

SELF-ASSESSMENT—EVALUATION PROCEDURES

The following lesson is designed for your use in determining a personalized profile of fitness. This profile consists of a series of contemporary evaluation exercises that require a minimum of time and materials and yet result in an acceptable degree of assessment accuracy.

Although a few of the exercises will be undertaken in class, the majority are left to you as work to enhance your understanding of physical fitness.

Specifically, the following exercises are found in this lesson: a Weight Range Appraisal Guide; a Cardiorespiratory Assessment consisting of a format relating to Heart-Rated exercise, the Cooper 12-Minute Test; and a dual directed Motor Ability Assessment consisting of flexibility and strength testing.

Weight Range Appraisal Guide

Although there currently exists no simple and accurate method of determining if a person is overweight, body type, total weight, and percentage of body fat are three criteria used to estimate abnormal body weight. Desirable body weight ranges are most accurately determined via specific body typing. Your body type can most easily be determined using a tape to measure the ankle girth at the smallest point above the ankle, with the tape snugged tightly. Refer your result to the frame chart below.[1]

	Small Frame	*Medium Frame*	*Large Frame*
Men	Less than 8 in.	8-9.25 in.	More than 9.25 in.
Women	Less than 7.5 in.	7.5-8.75 in.	More than 8.75 in.

Now consult the Table of Desirable Weights to determine an approximation of your acceptable weight range:

Metropolitan Life Insurance Company Table of
Desirable Weights (In Pounds) for Men and Women
of Ages 25 and Over (Indoor Clothing)

Height (with shoes on— 1-in. heels)	Small Frame	Medium Frame	Large Frame
MEN			
5 ft. 2 in.	112-120	118-129	126-141
5 ft. 3 in.	115-123	121-133	129-144
5 ft. 4 in.	118-126	124-136	132-148
5 ft. 5 in.	121-129	127-139	135-152
5 ft. 6 in.	124-133	130-143	138-156
5 ft. 7 in.	128-137	134-147	142-161
5 ft. 8 in.	132-141	138-152	147-166
5 ft. 9 in.	136-145	142-156	151-170
5 ft. 10 in.	140-150	146-160	155-174
5 ft. 11 in.	144-154	150-165	159-179
6 ft. 0 in.	148-158	154-170	164-184
6 ft. 1 in.	152-162	158-175	168-189
6 ft. 2 in.	156-167	162-180	173-194
6 ft. 3 in.	160-171	167-185	178-199
6 ft. 4 in.	164-175	172-190	182-204

Height (with shoes on— 1-in. heels)	Small Frame	Medium Frame	Large Frame
WOMEN			
4 ft. 10 in.	92-98	96-107	104-119
4 ft. 11 in.	94-101	98-110	106-122
5 ft. 0 in.	96-104	101-113	109-125
5 ft. 1 in.	99-107	104-116	112-128
5 ft. 2 in.	102-110	107-119	115-131
5 ft. 3 in.	105-113	110-112	118-134
5 ft. 4 in.	108-116	113-126	121-138
5 ft. 5 in.	111-119	116-130	125-142
5 ft. 6 in.	114-123	120-135	129-146
5 ft. 7 in. 133-150	118-127	124	139
5 ft. 8 in.	122-131	128-143	137-154
5 ft. 9 in.	126-135	132-147	141-158
5 ft. 10 in.	130-140	136-151	145-163
5 ft. 11 in.	134-144	140-155	149-168
6 ft. 0 in.	138-148	144-159	153-173

Courtesy of the Metropolitan Life Insurance Company
NOTE: For those between 18 and 25, subtract 1 lb. for each year under 25.

Remember, this table is only a guide and; therefore, should initiate your inquiry into more specific factors of your body such as weight management programs, weight reduction techniques, or calibration of fat and muscle tissue.

Police Officer Standards for Resting Heart Rate*[3]

FITNESS CATEGORY	AGE GROUPS		
	20-29 yrs (n = 88) Heart Rate (beats/min.)	30-39 yrs (n = 85) Heart Rate (beats/min.)	40-52 yrs (n = 30) Heart Rate (beats/min.)
Excellent	44 and below	44 and below	48 and below
Good	45 to 58	45 to 61	49 to 62
Average	59 to 66	62 to 69	63 to 69
Below Average	67 to 69	70 to 85	70 to 83
Poor	80 and above	86 and above	84 and above

*Measured 10 minutes after being seated in a quiet room.
(Females add 3-4 beats to Heart Rate levels.)

In general, the lower resting heart rate, the healthier the individual. In fact, the mortality rate for men and women with pulse rates over 92 is four times greater than for those with pulse rates less than 67.

Regardless of your positioning on the chart, remember that exercise (exertion) strengthens the heart so it can operate more efficiently.

Heart-rated exercise; however, is based upon the theory that this conditioning of the heart must relate to muscle overload. Training effects ensue only if the amount of the overload (training stimulus) is greater than some minimal amount of work necessary to maintain normal physiological function and that the level of this stimulus varies among individuals.

Due to this variance, research has defined the safe threshold level to produce measurable gains in function capacity at 60% of one's maximum heart rate range for your use.

Estimated Maximum Heart Rate (MHR) by Age[4]

AGE	15-19	20-24	25-29	30-34	35-39	40-44	45-49	50-54	55-60
MHR	205	200	195	190	185	180	175	170	165

Targeting of your heart rate for training can best be achieved by following the procedure detailed below:

Predicted maximum heart rate _____
  (Use chart provided)

2. Resting heart rate = _____
  (Taken several times while seated in a quiet room. Use the carotid artery in the neck through the application of light pressure with the index and middle fingers, counting the pulsations for ten seconds and then multiplying this figure by six to find the number of beats per minute.)

3. Now subtract your lowest resting heart rate from the predicted maximum heart rate. This figure now becomes your maximum heart range. _____

4. Finally, to determine your target rate for training define the intensity of the exercise per your perceived fitness level, i.e., 60%, 70%, 80%, etc. Multiply the percentage times your maximum heart range and lastly add your resting heart rate. _____

Cooper Test

For the past several years, running has become an integral part of the programming in physical fitness at the Institute. In our opinion, running is the best exercise activity since it allows for relative ease in overall skill attainment, allows for group participation, has validity in both short- and long-term law enforcement application, and unquestionably, it has precipitated the greatest amount of contemporary research.

New officers are encouraged vigorously to participate in an aerobic-based curriculum closely analogous to the popular design formulated by Dr. Kenneth Cooper, author of the books, *The Aerobics Way* and *The New Aerobics*.

Aerobics according to the dictionary is defined as "living, active, as occurring in the presence of oxygen." The term aerobic in its adaptation refers to the type of metabolic interplay utilizing oxygen in the production of energy for the body. Aerobics is then a program of numerous endurance exercises which require sustained effort. These types of exercise improve the heart, lungs, and blood vessels; the cardiorespiratory system. In totality, the Cooper program is designed for general population adaptation via a point system based upon the energy cost of the activity, i.e., the amount of oxygen utilized. For your future information, Dr. Cooper suggested that a minimum of 30 points per week is necessary in order to maintain satisfactory condition. The elaborate exercise/conditioning program charts unique to the Cooper continuous progress regimen have been included as part of the handout material in your physical conditioning classes.

Due to the limited nature of our training segment in physical fitness/personal defense, we have to assume that the recruit officer has met the level of functional fitness already as evidenced by his selection to a specific police agency. If because of a void in your department's screening you have not been medically cleared by a physician, you should limit your participation in the Cooper 12-minute walk/run test.

At the introductory stage of each training session, we may provide a testing medium to determine developmental and maintenance categories for both class ranking and overall fitness research. This test methodology consists of the participant's effort to cover the greatest distance in a fixed period of 12 minutes. Extensive research for over a decade has revealed a strong correlation between oxygen consumption, aerobic capacity, and has allowed us to maintain a constant measuring instrument of individual fitness through this simple running sequence. We shall use exclusively the charts below developed by Dr. Cooper which incorporates fitness categories, distance parameters, and appropriate age group adjustment.

Perhaps the closest paralleling population, researched by Cooper relating to police officers, would be various age groups within the United States military. The following charts are applicable to a "broad spectrum of the population" and now serve as a valid research foundational reference for specific police fitness profiles. You will be asked to become familiar with the defined fitness categories as appropriate to your age, and provide input during actual data-gathering sessions.

Whatever performance level you attain initially and as you further participate in our program design, let it serve as a catalyst for long-term patterns of fitness.

12-Minute Walking/Running Test[5]
Distance (Miles) Covered in 12 Minutes

Fitness Category		13-19	20-29	30-39	40-49	50-59	60+
I. Very Poor	(men)	<1.30*	<1.22	<1.18	<1.14	<1.03	<.87
	(women)	<1.0	<.96	<.94	<.88	<.84	<.78
II. Poor	(men)	1.30-1.37	1.22-1.31	1.18-1.30	1.14-1.24	1.03-1.16	.87-1.02
	(women)	1.00-1.18	.96-1.11	.95-1.05	.88-.98	.84-.93	.78-.86
III. Fair	(men)	1.38-1.56	1.32-1.49	1.31-1.45	1.25-1.39	1.17-1.30	1.03-1.20
	(women)	1.19-1.29	1.12-1.22	1.06-1.18	.99-1.11	.94-1.05	.87-.98
IV. Good	(men)	1.57-1.72	1.50-1.64	1.46-1.56	1.40-1.53	1.31-1.44	1.21-1.32
	(women)	1.30-1.43	1.23-1.34	1.19-1.29	1.12-1.24	1.06-1.18	.99-1.09
V. Excellent	(men)	1.73-1.86	1.65-1.76	1.57-1.69	1.54-1.65	1.45-1.58	1.33-1.55
	(women)	1.44-1.51	1.35-1.45	1.30-1.39	1.25-1.34	1.19-1.30	1.10-1.18
VI. Superior	(men)	>1.87	>1.77	>1.70	>1.66	>1.59	>1.56
	(women)	>1.52	>1.46	>1.40	>1.35	>1.31	>1.19

*< Means "less than"; > means "more than."

1.5 Mile Run Test
Time (Minutes)

Fitness Category		13-19	20-29	Age (years) 30-39	40-49	50-59	60+
I. Very Poor	(men)	<15.31*	<16.01	<16.31	<17.31	<19.01	<20.01
	(women)	<18.31	<19.01	<19.31	<20.01	<20.31	>21.01
II. Poor	(men)	12.11-15.30	14.01-16.00	14.44-16.30	15.36-17.30	17.01-19.00	19.01-20.00
	(women)	16.55-18.30	18.31-19.00	19.01-19.30	19.31-20.00	20.01-20.30	21.00-21.31
III. Fair	(men)	10.49-12.10	12.01-14.00	12.31-14.45	13.01-15.35	14.31-17.00	16.16-19.00
	(women)	14.31-16.54	15.55-18.30	16.31-19.00	17.31-19.30	19.01-20.00	19.31-20.30
IV. Good	(men)	9.41-10.48	10.46-12.00	11.01-12.30	11.31-13.00	12.31-14.30	14.00-16.15
	(women)	12.30-14.30	13.31-15.54	14.31-16.30	15.56-17.30	16.31-19.00	17.31-19.30
V. Excellent	(men)	8.37- 9.40	9.45-10.45	10.00-11.00	10.30-11.30	11.00-12.30	11.15-13.59
	(women)	11.50-12.29	12.30-13.30	13.00-14.30	13.45-15.55	14.30-16.30	16.30-17.30
VI. Superior	(men)	< 8.37	< 9.45	<10.00	<10.30	<11.00	<11.15
	(women)	<11.50	<12.30	<13.00	<13.45	<14.30	<16.30

*< Means "less than"; > means "more than."

Motor Ability Assessment

As noted earlier, Kaminski's second area which needed to be addressed by physical fitness progrmas for police was various aspects of motor ability.

Because of the widespread occurrence of cases of low back pain among officers, flexibility potential is a component of a total fitness assessment. Flexibility is the ability to move joints in the fullest range of motion. No general flexibility test measures the flexibility of all joints; however, the trunk flexion or the sit and reach test serves as an important measure of hip and back flexibility.

Sit on the floor with legs extended at right angles to a taped line or box. The heels should be spaced eight inches apart in line with the near edge of the tape or box. A yardstick is placed between the legs of the subject and rests on the floor with the 15-inch mark on the edge of the box. After several slow warm-up stretches, the subject should reach forward with both hands and record the distance achieved by the fingertips on the yardstick. The chart below lists police officer standards of flexibility for your reference:

Police Officer Standards for Flexibility[7]

FITNESS CATEGORY	AGE GROUPS		
	20-29 yrs (n=81) Flexibility (in)	30-39 yrs (n=84) Flexibility (in)	40-52 yrs (n=38) Flexibility (in)
Excellent	29 and above	26.4 and above	23.3 and above
Good	19.7 to 25.8	19.2 to 26.3	16.3 to 23.2
Average	16.6 to 19.6	15.6 to 19.1	12.8 to 16.2
Below Average	10.5 to 16.5	8.4 to 15.5	5.7 to 12.7
Poor	10.4 and below	8.3 and below	5.6 and below

Strength is the ability to exert force. Total body muscle strength is difficult to measure due to a vast array of muscle groups but some simple and accurate measurement is possible. A dynamic strength test through the full range of motion which correlates well with a total body strength criterion is the one repetition maximum bench press. These procedures are recommended for the safe and satisfactory strength testing effort.

1. Load the weights to approximately two-thirds of the person's estimated maximum weight. (Consult the charts provided for estimation basis.)
2. The subject should bench press this weight as a warm-up exercise.

3. Now begin to increase the weight increments 5-10 lbs. per lift; with the person usually reaching his or her maximum lift at the fifth, sixth, or seventh effort. The maximum lift in one repetition is scored. Consult the two charts for your peer ranking.

Police Officer Standards for One-Repetition Maximum Bench Press[8]

FITNESS CATEGORY	AGE GROUPS		
	20-29 yrs (n = 81) Bench Press (lb.)	30-39 yrs (n = 83) Bench Press (lb.)	40-52 yrs (n = 28) Bench Press (lb.)
Excellent	227 and above	201 and above	188 and above
Good	174 to 226	161 to 200	150 to 187
Average	147 to 173	141 to 1160	132 to 149
Below Average	94 to 1146	100 to 140	95 to 131
Poor	93 and below	99 and below	94 and below

One Repetition Maximum Bench Press Norms for College-Aged Men[9]
Body Weight Classifications (lbs.)

Fitness Category	120-129	130-139	140-149	150-159	160-169	170-179	180-189	190-above
				Pounds Lifted				
Excellent	170-150	175-155	185-165	195-175	205-185	215-195	225-205	235-215
Good	145-130	150-135	160-145	170-155	180-165	190-175	200-185	210-195
Average	125-110	130-115	140-125	150-135	160-145	170-155	180-165	190-175
Below Average	105-90	110-95	120-105	130-115	140-125	150-135	160-145	170-155
Poor	85-80	90-75	100-85	110-95	120-105	130-115	140-125	150-135

Muscular endurance is the ability to exert force over a prolonged period. Low levels of muscular endurance indicate inefficiency of movement and a low capacity to perform work. Two tests are provided for muscular endurance; the standard push-up and timed sit-up.

During the performance of the standard push-up, the officer should keep his back straight during the raising and lowering patterns. The chest should either touch or come within three inches off the floor. Again, a chart is provided for your comparison.

Police Officer Standards for the Pushup Test[10]

FITNESS CATEGORY	AGE GROUPS		
	20-29 yrs (n = 79) Pushups (repetitions)	30-39 yrs (n = 83) Pushups (repetitions)	40-52 yrs (n = 28) Pushups (repetitions)
Excellent	43 and above	37 and above	28 and above
Good	28 to 42	23 to 36	18 to 27
Average	20 to 27	17 to 22	13 to 17
Below Average	5 to 19	3 to 16	2 to 27
Poor	4 and below	2 and below	1 and below

The sit-up testing should begin with the officer lying on his back, knees bent, and hands clasped behind the neck. It is preferred that the feet be held down or under an object that can provide some degree of stabilization. The performance should include sitting up until the elbows make contact with the knees and then reclining back to a full lying position. Refer to the chart for comparison.

Police Officer Standards for the One-Minute Situp Test[14]

FITNESS CATEGORY	AGE GROUPS		
	20-29 yrs (n=81) Situps (reps/min)	30-39 yrs (n=83) Situps (reps/min)	40-52 yrs (n=28) Situps (reps/min)
Excellent	51 and above	45 and above	39 and above
Good	40 to 50	34 to 44	26 to 38
Average	35 to 39	29 to 33	19 to 25
Below Average	24 to 34	18 to 28	6 to 18
Poor	23 and below	17 and below	5 and below

The State of Illinois has established physical fitness training standards for entering any of the Illinois certified basic training academies and may provide an additional reference resource for comparison.

TEST 1

Threshold Weight

This is the weight that has been determined as the weight necessary to perform police tasks without undue effort and to minimize health problems due to overfatness. The score is pounds per height in inches.

Ht/In	Threshold Weight	Ht/In	Threshold Weight	Ht/In	Threshold Weight
52	75	63	134	74	217
53	80	64	141	75	226
54	85	65	147	76	235
55	89	66	154	77	245
56	94	67	161	78	255
57	99	68	168	79	265
58	105	69	176	80	275
59	110	70	184	81	285
60	116	71	192	82	297
61	121	72	200	83	307
62	128	73	209	84	318

1. % Fat

Those individuals who do not meet the threshold weight a % fat test is additionally administered. This is the percentage of body fat that has been determined as the level of overfatness that poses a health risk. The measurement is made with a skinfold caliper at several selected body sites. The final score is in a fat percentage.

	Male Age				Female Age			
Test	20-29	30-39	40-49	50-59	20-29	30-39	40-49	50-59
Percent Body Fat	20.4	23.5	25.5	27.1	27.7	28.9	32.1	35.6

2. Sit and Reach Test

This is a measure of the flexibility of the lower back and upper leg area. The test involves stretching out to touch the toes or beyond with extended arms from the sitting position. The score is in inches reached on a yardstick or other measuring device with 15 inches being at the toes.

	Male Age				Female Age			
Test	20-29	30-39	40-49	50-59	20-29	30-39	40-49	50-59
Sit & Reach	16.0	15.0	13.8	12.8	18.8	17.8	16.8	16.3

3. One Minute Sit Up Test

This is a measure of the muscular endurance of the abdominal muscles. The score is based upon the number of bent leg situps performed in one minute.

	Male Age				Female Age			
Test	20-29	30-39	40-49	50-59	20-29	30-39	40-49	50-59
1 Minute Sit Up	37	34	28	23	31	24	19	13

4. One Repetition Maximum Bench Press

This is a maximum weight pushed from the bench press position and measures the amount of force the upper body can potentially generate.

	Male Age				Female Age			
Test	20-29	30-39	40-49	50-59	20-29	30-39	40-49	50-59
Maximum Bench Press Ratio	.98	.87	.79	.70	.58	.52	.49	.43

5. 1.5 Mile Run

This is a timed run to measure the heart and vascular system's capability to transport oxygen. The score is measured in minutes and seconds.

	Male Age				Female Age			
Test	20-29	30-39	40-49	50-59	20-29	30-39	40-49	50-59
1.5 Mile Run	13:46	14:31	15:24	16:21	16:21	16:52	17:53	18:44

Footnotes for Chapter 4

[1]Johnson, P. B. "So You Really Want to Lose Weight?" Toledo: The University of Toledo, 1972.

[2]Kaminski, J. J. "Police Physical Fitness: A Personal Matter," *Police Chief,* April, 1975, pg. 39-40.

[3]Price, Clifford S., Pollock, Michael L., Gettman, Larry R., and Kent, Deborah A. *Physical Fitness Programs for Law Enforcement Officers: A Manual for Police Administrators,* Washington: U.S. Government Printing Office, 1977, pg. 244.

[4]Miller, David K. and Allen, T. Earl. *Fitness: A Lifetime Commitment, Minneapolis: Burgess Publishing, 1979, pg. 29.*

[5]*Cooper, K. H. The Aerobics Program for Total Well-Being,* New York: M. Evans and Company, Inc. and Bantam Books Inc., 1983.

[6]Cooper, K. H. *The Aerobics Program for Total Well-Being,* New York: M. Evans and Company, Inc. and Bantam Books Inc., 1983.

[7]Myers, C. R., L. A. Golding and Sinning, W. E. *The Y's Way to Physical Fitness,* Emmous: Rodale Press, 1973.

[8]Ibid.

[9]Ibid.

[10]Ibid.

[11]Ibid.

[12]Illinois Local Governmental Law Enforcement Officers Training Board. *Training Standard Manual,* July 1987.

Chapter 5

EXERCISE ROUTINES (CALISTHENICS)

The following exercise routines are provided for two reasons: (1) they serve as the initial warm-up sequence for each of our training sessions, and (2) these exercises can potentially develop muscular strength, maintain endurance capabilities, and provide essential flexibility for most of the muscle groups outlined.

Each routine will consist of an explanation inclusive of the described starting positions, purpose, movement, and suggested repetitions.

Neck Roll

Purpose: Gradually stretch muscles of the neck.

Position: Stand hands on hips, feet shoulder width apart.

Movement: Roll the neck slowly forward, sideward, and backward in a clockwise manner.

Repetitions: 10

Reverse this same movement and duration in a counterclockwise motion.

Repetitions: 10

Figure 5-1

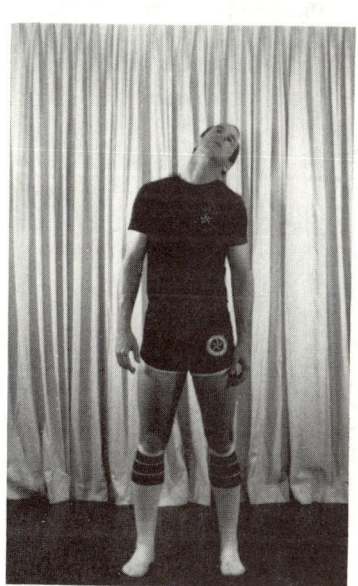

Figure 5-2

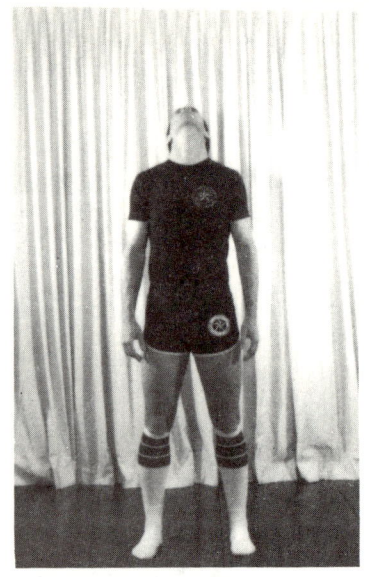

Figure 5-3   Figure 5-4

Arm Circles/Toe Raises

Purpose: To stretch muscles of the shoulder girdle and strengthen calf muscles.

Position: Stand with feet flat on floor, toes pointed forward. Arms are held at the side in a relaxed position.

Movement: Upon initiating the exercise, slowly bring your arms upward making large circles. As the arms are raised and crossed overhead, rise in a controlled motion on your toes.

Repetitions: 15 to 25

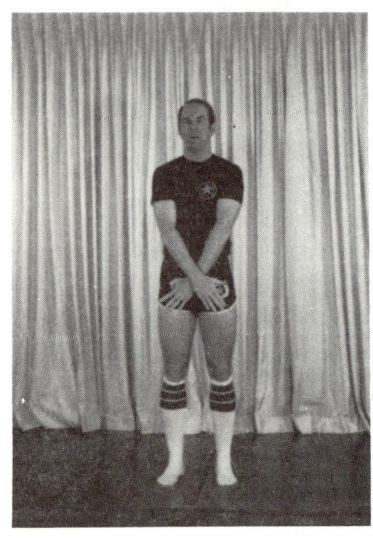  

Figure 5-5   Figure 5-6   Figure 5-7

        Figure 5-8   Figure 5-9   Figure 5-10

Hamstring Stretch

Purpose: To stretch muscles in the posterior leg and thigh.

Position: Sit on floor with one leg extended while the opposite leg is bent inward, so as to allow the foot to rest against the inner thigh of the extended leg.

Movement: Slowly bend forward attempting to touch the head to the knee. *Do not bounce into the forward bend.* Hold the stretch for 30 to 60 seconds. Repeat sequence with alternate leg.

Repetitions: 3 to 5

        Figure 5-11   Figure 5-12   Figure 5-13

Sit-Ups

Purpose: To strengthen abdominal muscles.

Position: Lie on back with knees bent and hands clasped behind head.

Movement: Slowly raise upper torso to an upright position and slowly return to the starting position. Each sequence of raise and return should require 3 to 4 seconds. Roll up into position avoiding a lower back snapping technique.

Repetitions: 25 to 35

Figure 5-14

Figure 5-15

Figure 5-16

Figure 5-17

Figure 5-18

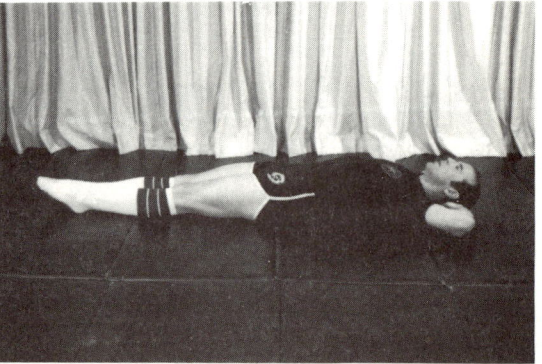

Figure 5-19

Lower Back/Abdominal/Hamstring Stretch

Purpose: To stretch and strengthen muscles in the area of the lower back, abdomen and posterior leg.

Position: Lie on back with legs suspended six inches off the floor.

Movement: Slowly lift the legs, bending the knees in the process. With the knees kept together, bring the straightened legs the rest of the way over the head to a point approximately 6" to 10" from the floor. Slowly return the legs back down to the suspended position.

Repetitions: 5 to 10

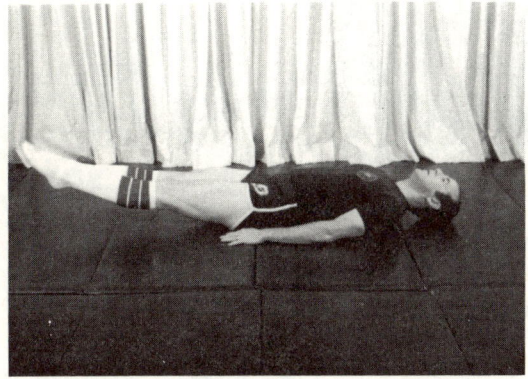

Figure 5-20

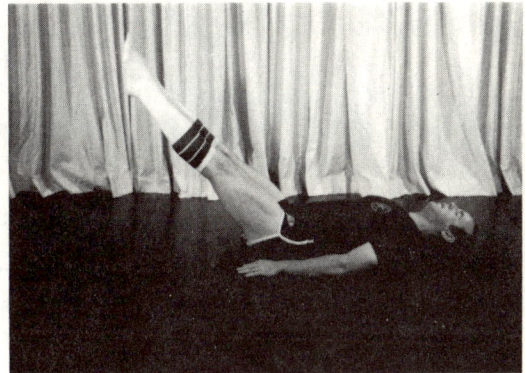

Figure 5-21

Figure 5-22

Figure 5-23

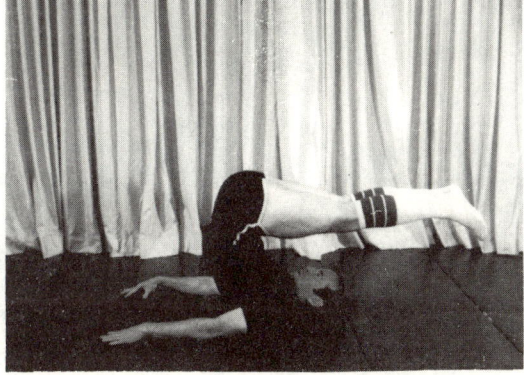

Figure 5/24

Push-Ups

Purpose: Strengthen the chest, anterior shoulder, and posterior upper arm muscles.

Position: Lie on stomach, body straight, hands flat on floor directly beneath the shoulders.

Figure 5-25

Movement: Push entire body upward while continuing to keep the toes touching the floor. Continue this upward motion until both arms are straight; lower body back downward until chin and chest barely touch the floor. Keep the head up and body straight.

Repetitions: Progress 20 to 30

Figure 5-26

Figure 5-27

Figure 5-28

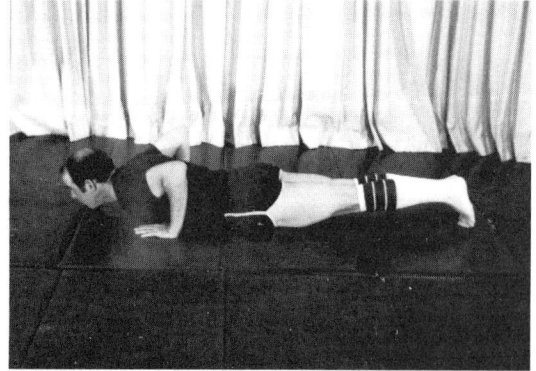

Figure 5-29

Windmill

Purpose: Stretch muscles of buttocks, abdomen, and posterior leg.

Position: Feet spread widely apart, knees can have a slight bend, trunk bent forward, arms stretched sideward.

Movement: 1) Touch right hand to left toe while left arm is moved backward and upward.
2) Touch left hand to right toe as right arm goes backward and upward.

Repetitions: Continue dual movement technique with rhythm to 25 times.

Figure 5-30

Figure 5-31

Figure 5-32

Figure 5-33

Figure 5-34

Leg Lifts

Purpose: Stretch buttocks, hip, abdominal and leg muscles.

Position: Lie on back, legs extended, knees locked.

Movement: Slowly lift both legs in unison and hold level approximately 6 inches off the floor for 20 seconds; then slowly lower to floor.

Repetitions: Commence the lift, hold and lowering sequence from 5 to 15. Hold-time with legs suspended can be increased from 20 to 45 seconds. Series of suspensions from 5 to 15.

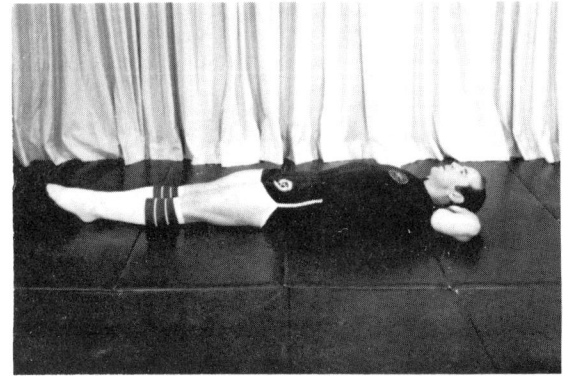

Figure 5-35

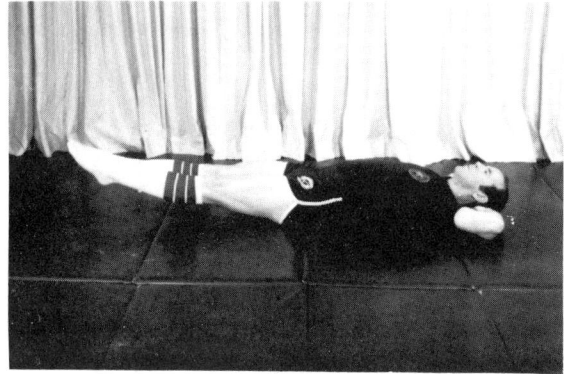

Figure 5-36

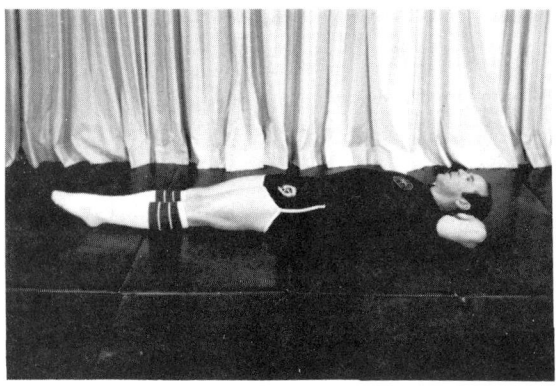

Figure 5-37

Groin Stretch

Purpose: To stretch upper leg and groin muscles.

Position: Sit in a relaxed manner with knees pointed outward; legs tucked, feet together.

Movement: Grasp ankles and slowly attempt to fold upper torso as close as possible toward the feet. *Do not bounce into this forward folding motion.* Once you have assumed the lowest forward tuck position, hold the stretch for 30 to 60 seconds.

Repetitions: 3 to 5

Figure 5-38

Figure 5-39

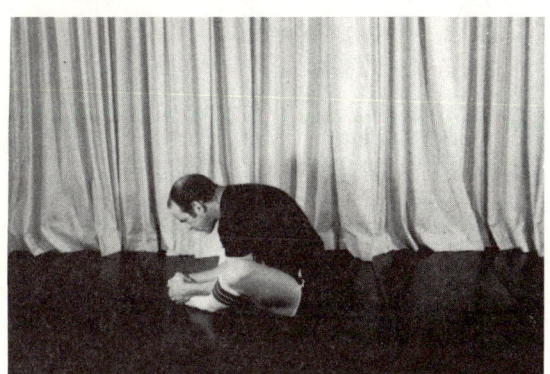

Figure 5-40

Figure 5-41

Ankle, Knee, Hip Rotation

Purpose: To stretch muscles in the area of the ankle, knee and hip.

Position: Stand in a balanced stance on one leg while lifting the opposite leg and commence the stretching sequence.

Movement: Beginning with the ankle, rotate the foot in small circles maximizing the full motion of the joint in a clockwise and then a counterclockwise motion. To exercise the knee joint, move the lower leg with full front to rear motion and side to side movement. Lastly, lift the entire leg making clockwise/counterclockwise arches with the hip. Repeat the sequence of motions with the alternate leg.

Repetitions: Repeat the contrasting movements for each joint area 10 to 15 times

ANKLE ROTATION

 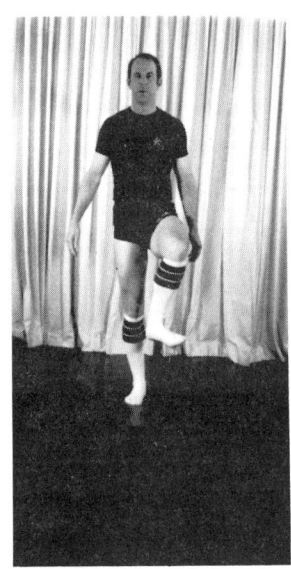 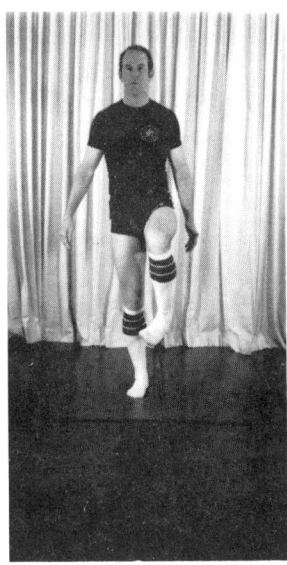 

Figure 5-42   Figure 5-43   Figure 5-44   Figure 5-45

KNEE ROTATION

Figure 5-46

Figure 5-47

Figure 5-48

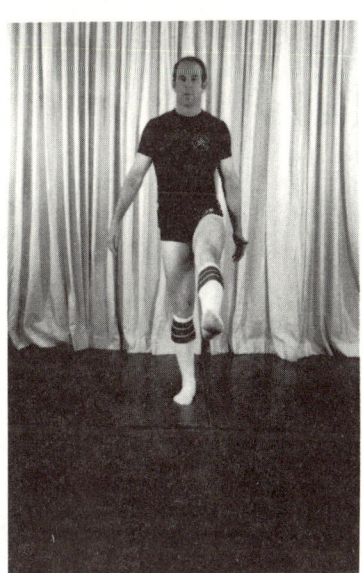

Figure 5-49

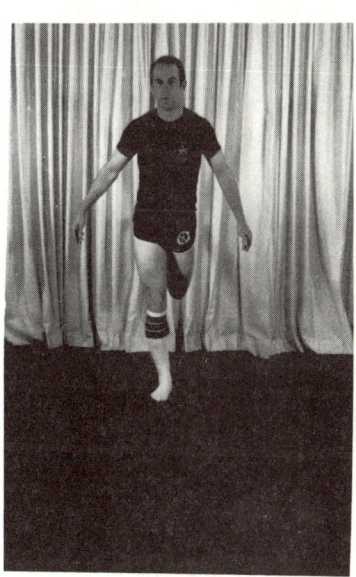

Figure 5-50

HIP ROTATION

Figure 5-51

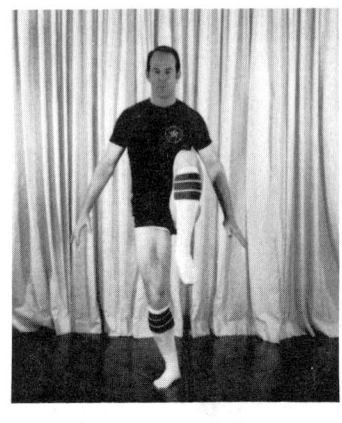

Figure 5-52

Figure 5-53

Figure 5-54

Figure 5-55

Wrist-Flex Series

Purpose: To stretch muscles in wrist region.

Position I: 1) Raise hand to be stretched in front of face, palm in. Place the thumb of the opposite hand between the little and ring fingers.

Figure 5-56   Figure 5-57

Movement: 1) Gently exert pressure via the thumb against the back of the hand while simultaneously moving the wrist being stretched toward the chest. Upon desired tension, hold the static position for 10 seconds, cease the pressure and allow the hand to return to the starting position. Repeat the process in repetitions noted.

Repeat exercise on alternate wrist.

Repetitions: 5 to 8 static conditions for 10 seconds each.

POSITION I

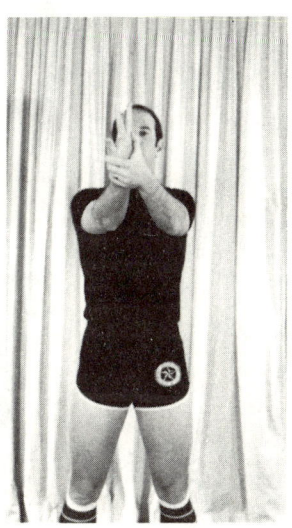

Figure 5-58   Figure 5-59

37

Wrist-Flex Series

Purpose: To stretch muscles in wrist region.

Position II: 1) Raise hand to be stretched in front of face, palm out. Grasp the ring and little finger with the opposite hand.

Movement: While maintaining a grasp on the fingers, rotate the stretched hand so that the palm remains outward while the edge of the hand moves toward the chest. Hold the static position for 10 seconds, cease the pressure and allow the hand to return to the starting position. Repeat the process in repetitions noted.

Repeat exercise on alternate wrist.

Repetitions: 5 to 8 static conditions for 10 seconds each.

Figure 5-60

Figure 5-61

Figure 5-62

Figure 5-63

Figure 5-64

Figure 5-65

Wrist-Flex Series

Purpose: To stretch muscles in wrist region.

Position III: 1) Raise the hand to be stretched to waist level with the palm down. Wrap the middle finger and thumb around the wrist joint with the palm of the same hand on the back of the grasped wrist.

Movement: With the grip established, begin the upward pulling movement of the affected hand from the waist level while simultaneously pushing down with the heel of the grasping hand. Once the hand being stretched has reached shoulder level, hold the static position for 10 seconds, cease the pressure and return it to the starting position. Repeat the process according to the repetitions noted.

Repeat exercise on alternate wrist.

Repetitions: 5 to 8 static conditions for 10 seconds each.

Figure 5-66

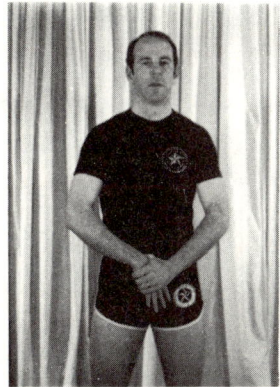

Figure 5-67

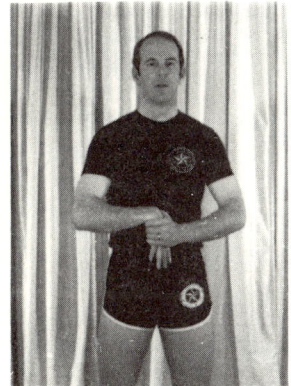

Figure 5-68

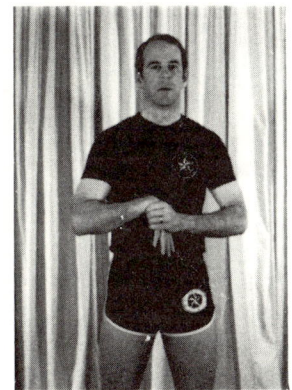

Figure 5-69

Figure 5-70

Chapter 6

UTILIZATION OF PERSONAL WEAPONS

It is reasonable to assume that circumstances may present an officer with a violent confrontation where regular issued weapons (i.e., baton, revolver, etc.)) are unacceptable. For example, the officer may be confronted while he is completely unarmed. He may be abruptly surprised, disallowing a time lapse of sufficient length to secure his weapon-based reaction; a drug crazed youth, a suicide prone female, etc.

The professionally sound and personally secure officer must be able to act effectively and efficiently in the use of natural body weapons. For many, this understanding and utilization requires an adaptation of a new totality of body perspecttive. Quite often, an individual conceptualizes his personal defense weapons as his fists or perhaps his feet. The limitations are a direct manifestation of the officer's exposure to the scope of tactical neutralization.

Initially, this perceptive limitation is replaced by conceptualizations that personal weapons include any and all parts of the body that have potential defensive utilization. The body is viewed as a defensive entity with both mental and physical parallels.

Of equal importance in eprsonal weapon utilization is the perspective of opponent vulnerabilities. These vulnerabilities must be recognized instinctively and become the targeting sight. Through practice and experience, you will better able yourself to perceive the unprotected areas of your opponent and initiate a defensive response. For instance, if an opponent's hands are in use attempting to effect a controlling hold, his body defenses have been decreased significantly and both the upper and lower torso may be vulnerable to a reactive strike. Remember, you won't have time to think of the proper response, it must be instinctive, reactive in nature. If the officer, while in a headlock, is trying to *think* of defensive options, he will pass out thinking. Instead, he should react with the immediate targeting of appropriate personal weapons to experienced sensed, vulnerable areas.

Personal weapon utilization must be correlated in terms of two dominant forces, balance and leverage. As mentioned earlier, balance is the most important asset in defending oneself or neutralizing another. Balance must be maintained beyond merely staying on your feet, it also includes totality of body control in reference to a positional advantage with your opponent.

Another important asset to the officer is the ability to use the force of leverage. Leverage involves the use of one's body as the means toward acquiring balance over the individual. Leverage may manifest itself in a variety of forms (i.e., throwing a subject *off* your back, application of a wrist technique, or the removal of a subject from a vehicle.)

One addititonal point concerning close contact or "in-fighting" techniques is the principle of mid-line targeting. Generally, it is suggested that blows are most effective if they are directed to the mid-line of the body both front and rear. Not only is this verticle perspective of attack paramount in effect, it is the body area usually made most vulnerable if a faking technique is incorporated by the officer.

In conclusion, use the utmost caution during the eventual practicing of each of these in-fighting techniques and their practical application within the work environment. Your perception of the situation, mirrored by your degree of force utilized, will be perhaps one of the most serious decisions you may exhibit within your career.

The following designations include some of the personal weapons available to the officer, with appropriate striking surfaces indicated.

Head—At least four thrusts with the head are defensively practical and these are illustrated in figures 6-1 through 6-4.

Figure 6-1
Forehead Thrust

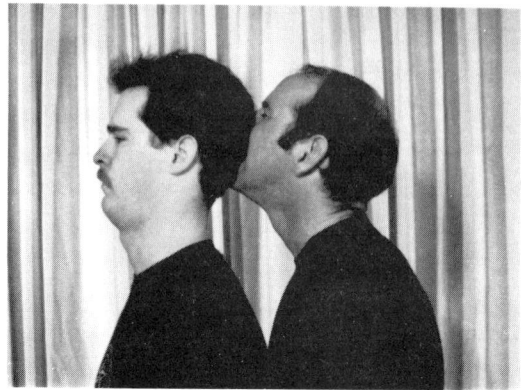

Figure 6-2
Backhead Thrust

Figure 6-3
Side Thrust

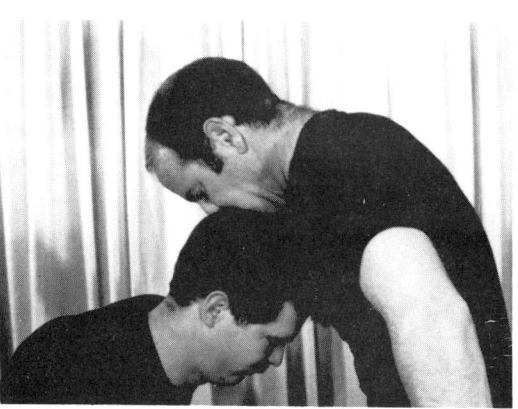

Figure 6-4
Upward Thrust

Hand Strikes—Although numerous hand techniques exist, nine have been selected and are illustrated in figures 6-5 through 6-13.

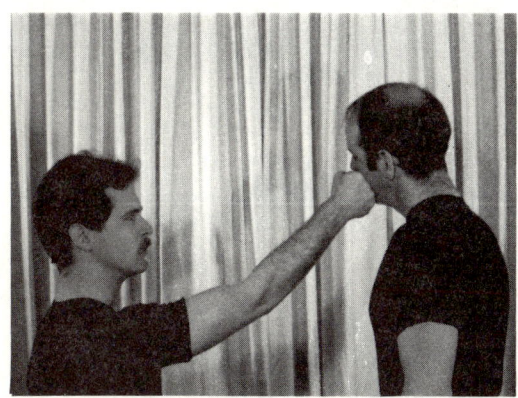

Figure 6-5
Forefist Strike

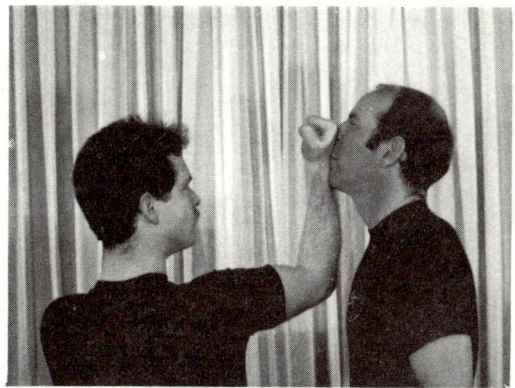

Figure 6-6
Backfist Strike

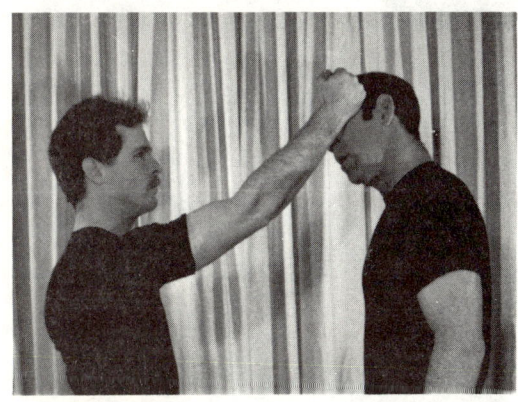

Figure 6-7
Hammerfist Strike

Figure 6-8
Palm Heel Strike

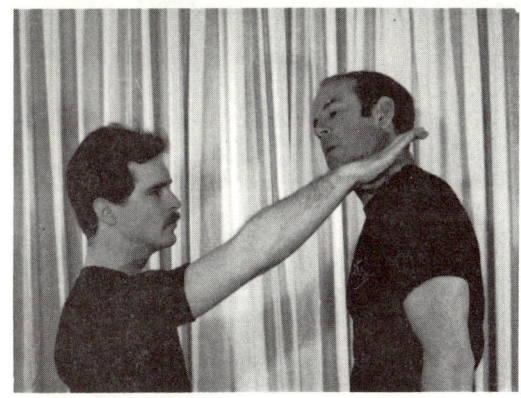

Figure 6-9
Ridge Hand Strike

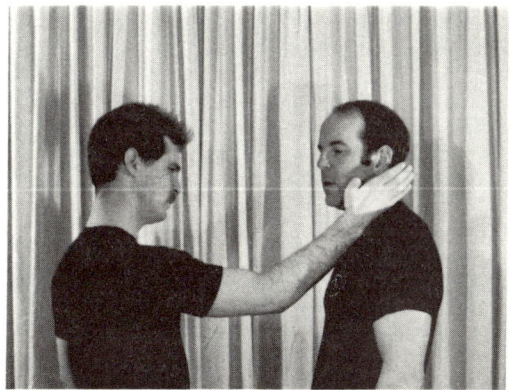

Figure 6-10
Knife Edge Strike

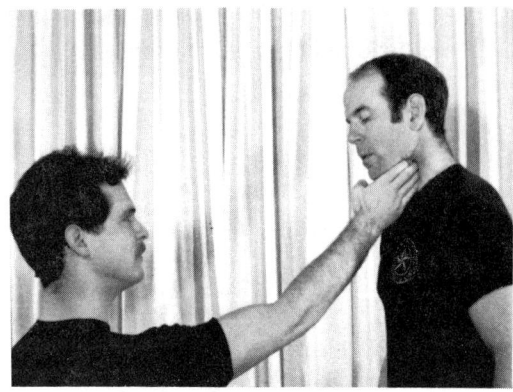

Figure 6-11
Spear Hand Strike
(Carotid

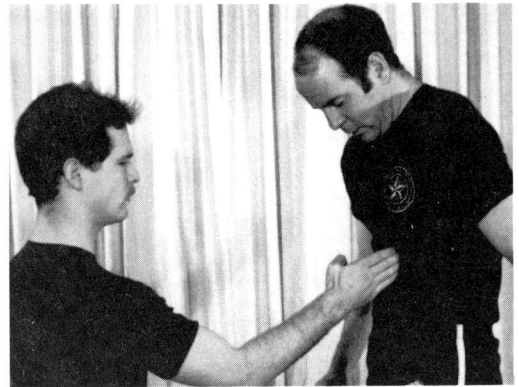

Figure 6-12
Spear Hand Strike
(Solar Plexus)

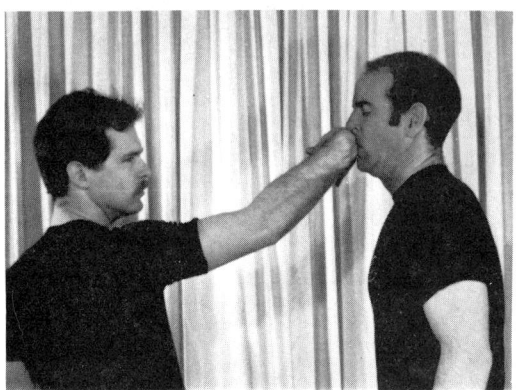

Figure 6-13
Back Wrist Strike

Elbow Strikes—The elbow can be a powerful weapon especially when delivered via either of the following techniques as illustrated in figures 6-14 through 6-16.

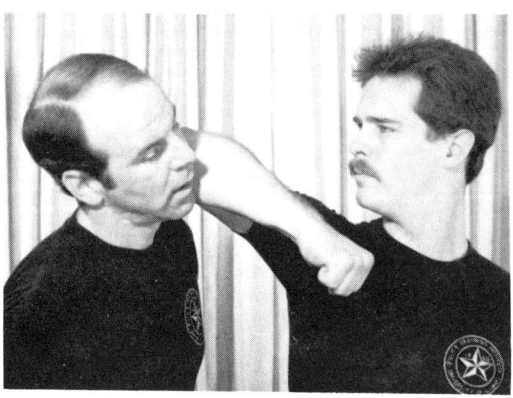

Figure 6-14
Horizontal Elbow Strike
(Forward)

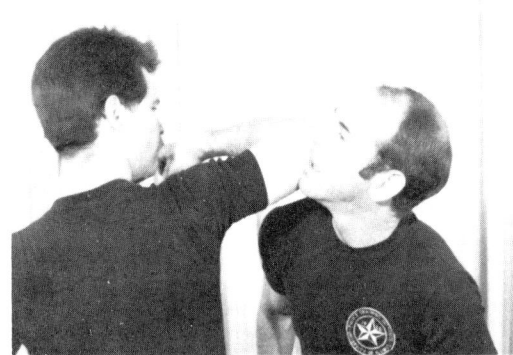

Figure 6-15
Horizontal Elbow Strike
(Reverse)

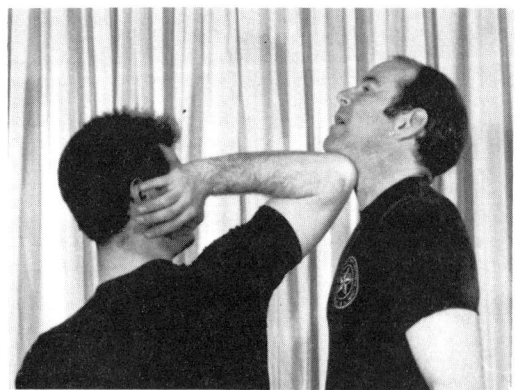

Figure 6-16
Vertical Elbow Strike

Knee Strikes—The knee also has various options of defensive capability as illustrated in figures 6-17 and 6-18.

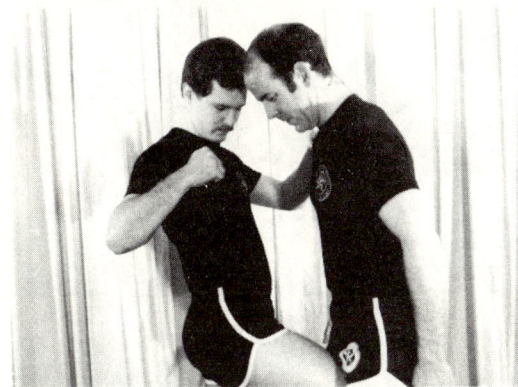

Figure 6-17
Vertical Lift Strike
(Groin)

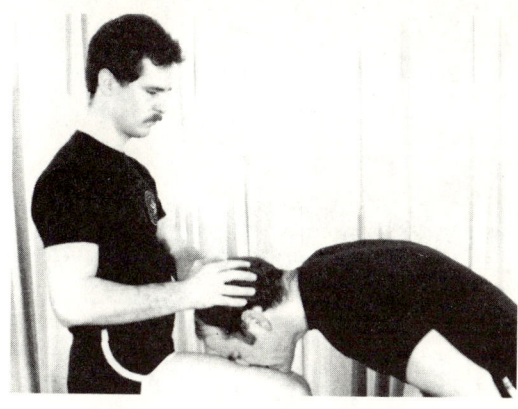

Figure 6-18
Vertical Lift Strike
(Face)

Foot Strikes—The foot too is a personal weapon effectively delivered in a technique mode as illustrated in figures 6-19 through 6-23.

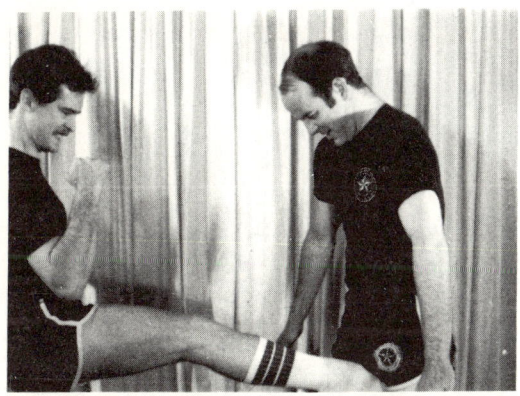

Figure 6-19
Top of Foot Strike

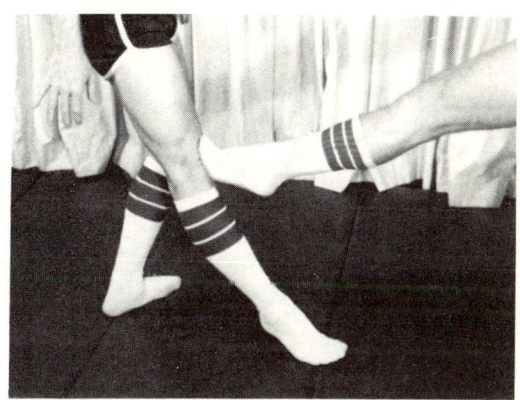

Figure 6-20
Ball of Foot Strike

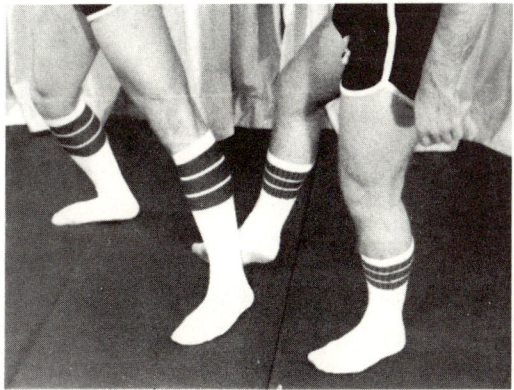

Figure 6-21
Instep Strike

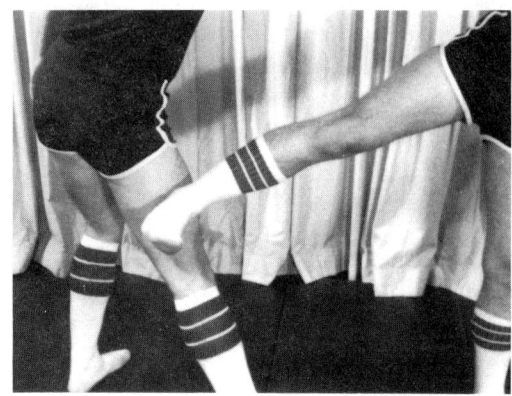

Figure 6-22
Heel Strike

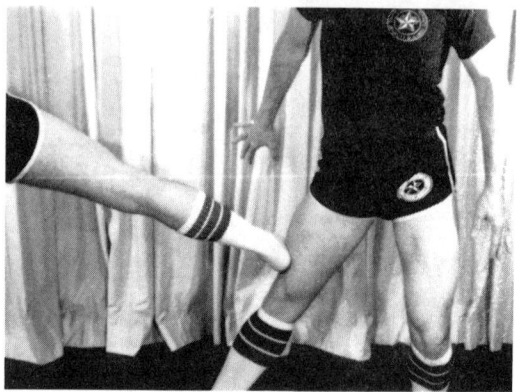

Figure 6-23
Knife Edge Strike

Targets of Vulnerability

The following anatomical sketches give numerical indications of vulnerable areas of the body. It must be noted that a strike to almost any of these areas could be fatal and therefore, their targeting should be undertaken in only the most extreme circumstances of jeopardy. The amount of force directed toward any of these areas fall within guidelines of law and liability. An asterisk is used to designate those areas of the body where a striking technique is most likely to result in a fatality.

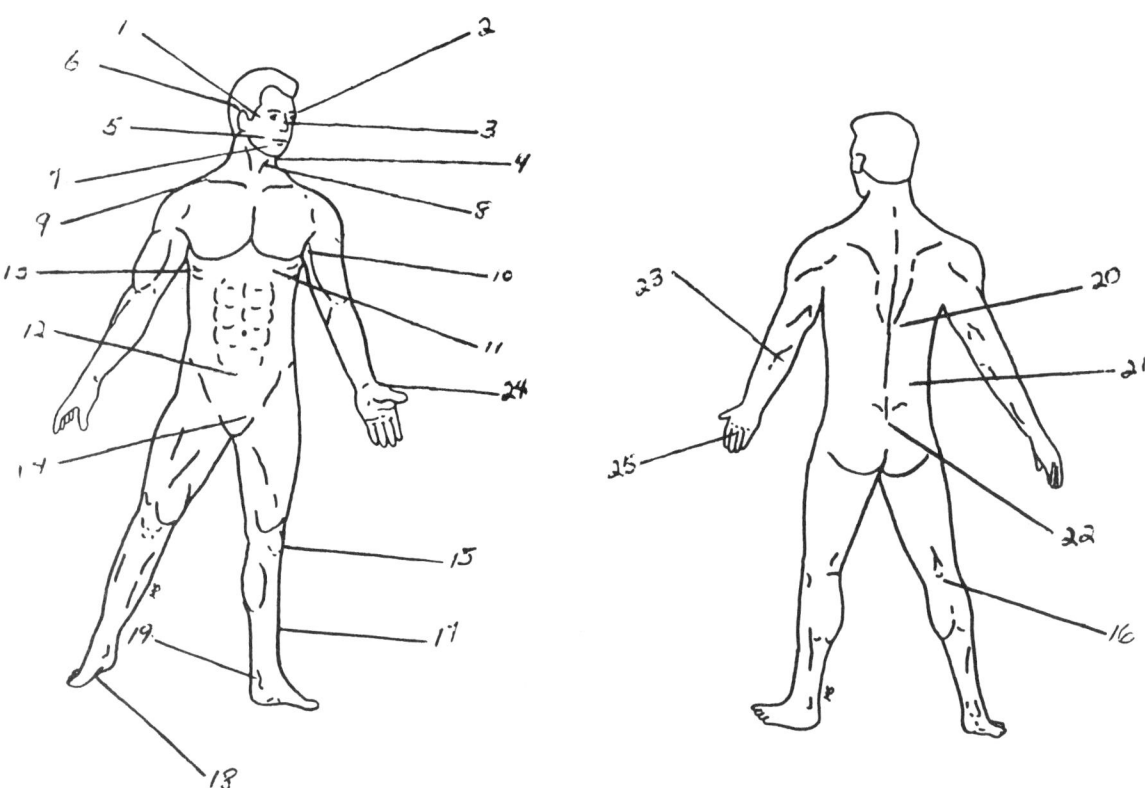

Figure 6-24

Each of the numerically listed vulnerable areas in the diagrams are given some narrative supplement both as to the most applicable personal weapons utilized and their corresponding physiological results.

*1. Temple—a strike with a fist, hand, knife edge or elbow could result in a reactive range from debilitation to death.

*2. Eyes—fingers thrust into the eyes could result in eye damage and/or death.

3. Bridge of nose—knife edge or heel of the hand can bring about severe pain and even temporary blindness.

4. Sides of the neck—hand, knife edge can result in unconciousness.

5. Lips—a rapid and powerful backfist results in severe pain.

6. Ears—a double palm compression to the ears can cause disorientation, inner ear damage, and the potential for severe concussion.

7. Chin—a palm heel strike can cause unconciousness or neck strain.

*8. Larynx—hand knife edge strike can bring about severe pain, respiratory difficulty, even death.

9. Collarbone—a knife edge blow or downward elbow strike can result in severe pain, fracturing, or complete debilitation of the appropriate arm.

10. Armpits—spearhand strike or fist into the area may cause severe pain or even temporary paralysis.

*11. Heart—fist or hand heel strike can result in death.

*12. Solar Plexius—fist, spearhand strike, or upward elbow strike can cause severe pain, respiratory impairment, or death.

13. Rib Cage—palm-heel strike, knife edge hand strike, heel or knife edge foot strike may result in a cracked rib, collapsing of a lung, or more extensive organ damage.

*14. Groin—an upward lift of the knee, foot strike, or hand slap and grasp can produce severe pain, tissue damage or even death.

15. Knees—a knife edge kick, heel strike or ball of foot strike to the knee can produce severe pain, debilitation or joint dislocation.

16. Back of Knee—a knife edge kick or heel strike can dislocate the knee joint as well as damaging the corresponding nerves in the area.

17. Shins—a knife edge kick, ball of the foot strike, or heel strike causes severe pain.

18. Insteps—a powerful foot stomp on the subject's instep can produce severe pain and fractures to the small bones of the foot.

19. Ankles—a knife edge kick or ball of the foot strike to the area results in severe, debilitating pain.

*20. Spinal Column—a knee lift, elbow strike or foot thrust technique directed in this area could cause paralysis or death.

*21. Kidneys—a knife edge strike, forefist thrust, knee lift, or foot strike can cause severe pain, tissue damage, shock effects, or even death.

22. Tailbone—a knee thrust or foot strike can result in severe pain, nerve damage or paralysis.

23. Elbows—open palm strikes, knife edge strikes or hand heel strikes to the elbow joint may result in severe pain, temporary debilitation or actual dislocation.

24. Wrists—a grasp of the attackers wrist and commensurate twisting, flexing, bending or pinching can cause severe pain and possible dislocation.

25. Fingers—a grasp of the attackers fingers and commensurate twisting, flexing, bending or pressure application produces severe pain and possible dislocation.

Chapter 7

CONFRONTATIONS: MENTAL PERSPECTIVES

In addition to the descriptive aspects of the personal defense techniques that are formulated in this manual, there exists a variety of concepts that must be addressed in order for the officer to further internalize the proper perspective regarding personal defense. Throughout this manual, various techniques will be stressed, with minimal emphasis on the mental perspectives that are conducive to successful personal defense. It is vital that you be exposed to these perspectives, and if possible, internalize these for use once on the street.

Tactical neutralization techniques are merely one-half of effective street survival. Physical involvement in a confrontation through defensive techniques is not an insured means to success. There exists a myriad of elements to consider in any confrontation that are outside the domain of strength, agility, balance, and reaction time. These aspects are found in the realm of the mind. Consequently, the officer must begin to realize that merely knowing, remembering, or attempting tactical neutralization techniques will neither diffuse an assaultive confrontation nor successfully guarantee your personal protection. *Mental* involvement is as necessary as the physical involvement.

Mental Involvement

Mental involvement in violent confrontations is the other half of the complete sphere of personal defense. It can be viewed as the sights of a weapon; the techniques . . . the ammunition. While the weapon may be fired without the sights, it is less effective and indeed, can result in tragic consequences. The same holds true for personal defense. Without mental involvement, the physical techniques are less effective and increasingly ladened with liabilities. Consequently, certain perspectives are vitally necessary to solidify your knowledge of the techniques you will be exposed to in your current training experience.

Fundamentally, you should realize that when you are involved in a violent confrontation, regardless of the circumstances, it is serious by virtue of its nature. Confrontations with police officers are serious merely because of the officer's involvement. The officer must realize that if an individual assaults a police officer, he is assaulting an institutionally armed individual. If an individual is willing to assault an armed police officer, one could contend that he would be in no way hesitant to assault an unarmed citizen. The fact that any assault of a police officer is an assault against an armed individual should indicate the severity of the assault. Simply put, the officer should realize that a violent assault should not be viewed with the same perspective as schoolyard fights. Prior to policing, your life experiences may interfere with your role identification. No longer are there rules for combat, no points, no referee overseeing the safety . . . no longer a sport, now it is survival.

Carrying this perspective one step further, personal protection should now take on a significant importance with the police officer. It should not be viewed with a sensationalist perspective or a lackadaisical attitude. Personal defense is as intense and as serious as the officer's personal motivation toward total professionalism. The proper perspective for learning personal defense does not lie with peer attitudes or departmental requirements. This responsibility lies with *you* . . . you determine, both mentally and physically, your ability to respond! If an officer has successfully achieved the tactical transition of the mind and body, the skill, and the psyche, he will not only experience street survival . . . more so, street success. Of the utmost importance, the fact that the officer realize his or her individual perspective or attitude adaptation concerning personal defense will dictate, more than any other variable, the capacity for learning. If there is no concern for learning, learning will not take place. If there is only a marginal concern for learning, only marginal requirements will be achieved. If there is a sincere desire to learn the art of protecting oneself, then anything that is said during the course of instruction will have some type of benefit to the officer's personal protection and controlling capacity.

Since an officer is involved with defending himself and others, he should realize that the seriousness of the confrontation is his responsibility. If an officer is assaulted and can successfully defend himself without sustaining seriuos injury to the attacker and himself, the seriousness of the attack is reduced. This is the essence of tactical neutralization. This occurs because the officer has controlled the situation before it has had an opportunity to get out of hand or redirected the forces or violence in a direction of diffusion. Conversely, if the officer cannot defend himself by any other means than deadly force, the seriousness of the incident is elevated to the ultimate extreme. Although we are dealing with extremes in these two examples, the point should be clear. In an occupation where personal defense is used more frequently than with any other, one readily available aspect of the officer's professionalism is indicated by his knowledge of neutralization techniques. As the officer has the responsibility for the use of his firearm, so too he has the responsibility and liability for his neutralization skills.

On a more pragmatic level, there are certain concepts that will aid in controlling a violent situation when the occasion arises. As we learned earlier, during times of crisis or duress, the human body's physiology alters significantly. There are increases in blood pressure, respiration, heart beat, and adrenaline secretion. The last of these increases is of the most importance. Adrenaline flow has been attributed to a varity of odd occurrences during times of duress. Stories of superhuman feats, such as successful two story plummets, lifting items two or three times the weight of an individual, previously unknown running speed and a variety of other events, are merely a few manifestations of the natural flow of adrenaline. Adrenaline is a form of a bodily defense mechanism that increases one's strength, among other effects. The adrenal gland can be used to assist or hinder an individual during times of crisis. If the adrenal strength is undirected and uncontrolled, it merely serves to exhaust an individual prior to his normal threshold of fatigue. That is, the individual will become tired prematurely in regard to the activity that is taking place. However, if used correctly, adrenaline will assist the individual in maintaining an increased level of strength and reaction time that could not normally occur. Consequently, the adrenaline flow that occurs during times of duress should be viewed as a potential ally.

During times of crisis, the officer must adopt a perspective that not only is applicable to situations of duress, but is applicable to all forms of activity. Economy of movement is directly related to the concept of hyperactivity. If the adrenaline that flows during duress can be harnessed, the officer is in a position of extreme benefit. Frequently, the adrenaline that flows during a violent confrontation is wasted on wild punches that are unfocused and wasteful movements of every conceivable type. The strength that was gained by virtue of the adrenaline flow is wasted on these types of activities. When the confrontation reaches a point where the officer gains control, the energy that was dispensed in the wasted activities drains the officer of all energy. Consequently, the officer has little strength left to control the situation. The officer should be aware enough of his physiology to control these types of wild and wasteful activities. If the officer can refrain from these activities and conserve the precious amount of adrenaline, then the adrenaline strength can be used on activities where and when it is needed. The term economy of movement refers to conservations of movement to save the energy gained by adrenaline. One has to consider the dispensing of energy as similar to the dispensing of ammunition. If one is given six bullets during an attack, one should not discharge all six rounds in an uncontrollable fury. The control of the attack may only necessitate one, well placed, round. The adrenaline use must be in the control of the officer.

Again, during times of crisis, the natural tendency for one's physiology is to increase significantly— the individual becomes hyperactive. This hyperactivity results in wasted energy; energy without direction. And because of this natural tendency to be wasteful, the fatigue threshold will occur prematurely.

During times of crisis and adrenaline flow, the single most significant element for the control of one's own physiology lies with respiration. The control of the bodily function of respiration will have a variety of beneficial effects. Primarily, if respiration is controlled, the individual will be less inclined to become "winded" than if the individual breathes uncontrollably. Furthermore, by controlling the breathing, the individual will relax himself and ease the tension that is a result of the adrenaline. The control of respiration will also serve to assist in the reduction of blood pressure, heartrate, and anxiety. Control is achieved by merely reducing the respiration rate to a lower frequency and by developing a cadence. The cadence should lower the breath rate and increase oxygen consumption (slow breathing down and take deeper breaths.) Once the cadence has been established, the rest of the body's processes will subside concurrently and with relative proportion. The officer must not resign himself to adopting a passive role as far as his bodily functions are concerned. This concept is analogous to a situation in which the officer is driving a squad car and another vehicle pulls out in front of him. If the officer passively lets the squad car continue as it is, a traffic crash will occur. If the officer attempts to control the vehicle, and "drives defensively," the crash is less likely to occur. The same can be said of the officer's physiological functions during an intervention in a crisis. If the officer lets his body's physiology rampage without any constraints or any attempts to control it, he will be a passive product of his physiology. If the officer places some types of control on his physiological processes, then he will be in better control of himself and consequently the situation.

Mental Sequencing

The officer should refrain from considering the techniques to be learned as a cause and effect type of situation. With a limited amount of knowledge in personal defense, there is a tendency for one to attempt to apply one technique to every type of personal defense situation. Not only is this illogical, it is also dangerous. Many times the officer will find himself forcing a technique that is inappropriate because he knows no others. If a technique can not be used, it should not be even attempted, let alone forced. There is no possibility of instruction of techniques for every type of tactical situation that arises. All that can be offered is a set of rules to follow with specific techniques that may or may not be applicable to the individual situation. In such situations, it behooves the officer to realize that he must rely on his knowledge of personal defense. Not only must he assess the situation, but he must determine which techniques are the most appropriate and then apply them. In the purest sense of the concept, the final examination for personal defense does not occur in the classroom, it occurs on the street. However, there are some specific guidelines that we can establish to assist the officer. One of these is the mental sequencing of techniques. Total mental sequencing of neutralization techniques is derived from two distinct domains. In the subjective domain, the officer must first assess the situation and deductively select technique orientations. Within the objective domain, technique specificity and tentative sequencing, as well as the compilation and assimilation of situational feedback, are imperative. Both domains structure effective execution of tactical neutralization.

The mental sequencing that should occur prior to an immediate confrontation should be exercised by the officer at all times. This "mental battle" is in the form of anticipating the subject's first move and establishing a countermove. It consists of a mental agenda of techniques. The officer should be conceptualizing his reaction to any possible assault or movement that may occur. This should not limit the officer to one technique. The officer should have a mental counter for every possible move by the subject. Not only should he have a mentally established, primary countermove, but he should place himself in position to affect the technique should the occasion arise. Secondly, merely pre-planning a single move in advance is like playing a chess game with a "checkers" perspective. The officer should have two or three techniques pre-planned to initiate if the primary technique proves ineffective. This will avoid attempting merely one technique and being both physically and mentally disoriented if it proves ineffective. Merely because a technique is appropriate does not mean it will be affected. Consequently, if a technique fails, for whatever reason, the officer should have established a sequence of techniques to resort to in his protective skills arsenal. This type of pre-planning or establishing the mental battle plan will hinder unexpected attacks from occurring, and consequently, assist in the defense of the officer.

Spatial Relationships

A very worthwhile perspective to consider is that of "personal space." This concept refers to the fact that once any object has neared an individual, there is a certain distance at which the individual becomes uncomfortable. This is usually associated with interpersonal communication. There are certain individuals that tend to violate the living space concept by standing or talking inches from the individual they are addressing. This concept applys to police officers because they will encounter a great deal of individuals that will violate this space. If the individual is hostile toward the police officer, this can have many grave repercussions. Consequently, the officer should be aware that no individual should violate the officer's "personal space." An imaginary sphere should enclose the officer and extend in every direction for approximately 3-4 feet. No individual should intrude on this space. If, for some reason, an individual attempts to violate this space, the officer should move in such a manner as to maintain this working space. This working space allows for an increased reaction time to any possible assault that may occur. The vital concept is that it is the *officer's* perrogative to violate the working space. He intrudes into a suspect's space to control more effecttively and is constantly vigilant to prevent any reversal of that spacial intrusion.

Execution Refinement

The final mental perspective that needs to be considered by the officer is one referred to as the "kiop." This Korean phrase is the nomenclature for the yell delivered by those involved in karate following an attack. This has mental benefits that should be considered by the officer and eventually adopted if desired. The kiop is used for a variety of purposes in personal defense of which the most significant is that of psychology. A scream during the delivery of an attack or counterattack has profound psychological repercussions for the subject. Not only does the yell indicate the control of the officer of the situation, it has a disturbing effect on the assailant. The kiop has karateka influences and has a limited indication of expertise. The beauty of the kiop is manifested in a situation where a subject attacks a police officer and attempts to strike the officer. The officer thwarts the attack and delivers a deterrent punch combined with a kiop that subdues the subject. The kiop in combination with the deterrent punch will reduce the tendency for future attacks. To merely thwart the attack without the kiop may antagonize the subject. Simply put, the kiop serves to influene the subject into cooperation. A corollary benefit of the kiop is that during the yell the diaphram and abdominal muscles are contracted which will reduce injury if the officer is struck. Consequently, a combination of physical and mental elements are fused.

The objective of each of the previous perspectives is for the officer to realize that violent confrontations involve more than a physical element of personal defense. Violent confrontations necessitate the officer's awareness of all elements, repercussions, actions and every factor of any potential personal defense situation. There exists a myriad of mental perspectives to consider and eventually adopt to enhance one's personal defense. However, this broad overview of fundamental perspectives should give the officer some type of indication as to the mentally sound path to successful personal defense. The proper mental perspective for learning and applying personal defense will strengthen the officer's confidence in himself and his ability to handle violent confrontations. With this confidence, the officer can begin to reduce the severity of future confrontations as well as the risk of injury to himself and others.

Chapter 8

STREET STANCES

Newsweek magazine once voiced that the most difficult occupation in a democratic society is that of being a police officer. This difficulty at no time takes greater expression, nor is more easily evidenced than during the period of the police/citizen confrontation.

Alert Stance

As accomplished as we may be via education or life experience, we can never totally predict the actions or reactions within a citizen encounter. Therefore, it is best to approach with caution, in a stance of constant protection, tempered with passivity in style and appearance.

It is believed that the "alert or interrogation stance" best exhibits those essential traits and should be incorporated through practice by the officer until its features become constant through instinct. The essence of the "alert stance," indeed the essence of combatives, in sport or on the street, is balance. Balance is best achieved by assuming a stance with a solid stable base and a constant realization of the body's center of gravity.

The "alert stance" places the officer at a safe distance (4-6 feet) to the front and right of the subject. From this position, the officer's chances of being struck or kicked are reduced substantially. The stance specifically places the officer's weak foot forward pointed toward the subject. The strong foot should be approximately a shoulder width behind the weak foot at approximately a 90° angle. This foot positioning allows you the greatest strength, balance and mobility. At the same time, your revolver is positioned the greatest distance away from the subject. (See Fig. 8-1)

Figure 8-1

The officer's upper torso must also mirror a status of safety as well as defensive countermoves. This dual capability is best realized by the officer keeping his elbows close to his ribs while keeping his hands high.

The officer should avoid the subject's eyes, thus avoiding distractions (fakes, hostile expression, etc.) and should place primary sight concentration on the suspect's upper sternum area. This "center mass" visual concentration allows the officer to detect punches from either arm and, likewise, kicks from either leg.

This stance, like balance itself, must be a natural product of the confrontation. Instinctively, your movements must be controlled, practiced exercises ever vigilant of a potential hostile threat.

Body Clock

A method of tactical assessment is the segmentation of a subject's body into numerical risk levels. For instance, if we were standing above a suspect, looking down, you could identify risk potential with clocklike description. For instance,

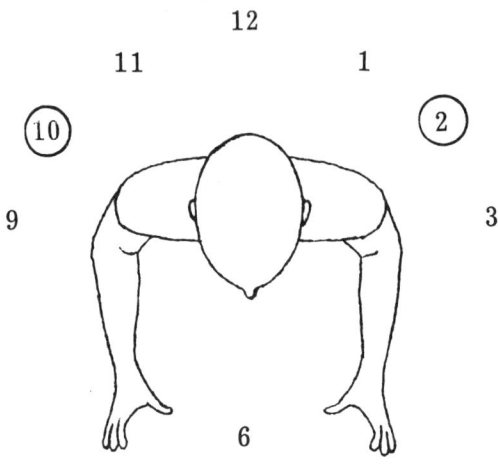

as we look at the drawing and subsequently assess the risk levels, one can see that the 12 o'clock, or rear position is the lowest on threshold level of risk. As we move from the rear position on around the body to the highest risk level, 6 o'clock, or front position, the variance in increased, vigilance is evident.

Our initial as well as on-going position should be ever cognizant of this "body clock" in our citizen contacts. Positioning preparation and planning must be flexible to the actions of the subject(s) and the enforcement environment. In most situations, the 2 or 10 o'clock positions are the most practical and efficient tactical approach foundations.

Arrest Stance

Once it becomes necessary to actually place the subject under arrest, the officer should move to an even more defensively advantageous position; the "arrest stance." The sophistication of this technique is critical so as to control this highly sensitized level of confrontation. You "talk" to the subject via your nonverbal cues; touch, spatial situation, or facial gestures. Loss of control within any of these areas can precipitate a violent escalation of the encounter.

Control is the core element of *your* intentions. If you merely are escorting the passive suspect to the squad car, give nonverbal cues of your intention; firm but not forceful grip, close but not crowded positioning, etc., along with supportive, diffusing dialog.

However, if you sense potential resistance from the subject, your control intentions should be clear initially. The grip is now a grasp for maintenance, for possible execution of higher levels of control; come-alongs, takedowns, etc. The arrest position is best achieved by moving smoothly but rapidly to the side and slightly to the rear of the subject's rear flank. Even more important now that you have initiated an offensive gesture toward the subject, is the elevation of your senses of observation and perception, alertness, and as always, the totality of balance.

As you maneuver to the rear/side position, remember to slide your lead foot toward the subject and quickly swing your rear leg into its final trailing stance. (See Fig. 8-2 and Fig. 8-3.)

Figure 8-2

Figure 8-3

The upper torso will correspond by reacting in quick, controlled body movements depending upon your rear/side position next to the subject. If you assume the arrest stance on the subject's right side, place your left hand on his elbow so as to avail yourself of potential pressure-point utilization as well as joint pronation and various takedown techniques. The right hand grasps securely the entire right wrist of the subject in such a manner that it is completely encircled. This hand positioning is again ideal for initiating numerous control, come-along, and takedown techniques noted later.

Protective Stance

On your initial contact with a subject, you could immediately be thrown into a position of physical jeopardy. This same state of physical jeopardy could occur throughout the sequence of interaction between you and your subject; the mild, passive, intoxicated driver becomes incensed when he sees his car being towed by a wrecker, the brother of the subject you have on a warrant believes family lines are stronger than any number of men in uniform, etc.

Because of these and other potential attack situations, it is felt that a protective stance should be employed to maximize your status of safety. Here, your lower torso assumes a "closed stance" with the strong leg to the rear and the front leg in a paralleling position of balance. The officer turns his weak side perpendicular to the attacker, granting the greatest distance from his weapon from the subject and allowing the least vulnerable areas of body exposure. The lower torso is also in an excellent position to respond with its own defensive armory of snap and thrust kicks.

The upper torso also must respond to the apparent attack. It too is turned to the side minimizing the exposure of vulnerable frontal areas. The weak hand is held to the side along the midline of the weak leg to deflect lower torso and midtorso strikes. The strong hand is held in readiness, face high, close to the body. Your own practice will profile the idiosyncracies unique to your adaptations (foot parallel distances, open or closed hand preference, etc.) necessary to present a strong, balanced protective stance. (See Fig. 8-4 and Fig. 8-5.)

Figure 8-4   Figure 8-5

PATTERNS OF CONFRONTATION

One Officer/Single Subject

For the most typical of street encounters, the lone officer may best approach a suspect from his right side on a path closely perpendicular with his presumed strong side. This placement best affords the officer time to react to a kick or punch. (See Fig. 8-6 and Fig. 8-7.)

Figure 8-6   Figure 8-7

Remember body positioning must be fluid and manipulative without the suspect realizing that he is being "jockeyed" into a more vulnerable position.

Two Officers/One Subject

Under this ideal situation, the solitary suspect is approached by two officers in a manner resembling a triangle. Each officer approaches the subject from either side with the triangle's apex, the suspect, being confronted.

This doubled approach splits the subject's attention and he, therefore, remains at a disadvantage. Maximum effectiveness of the triangle approach is realized when one of the officers serves the primary officer and totally directs the conversation with the subject from his prearranged front/right position. (See Fig. 8-8.) From this position, the secondary officer can potentially move to the suspect's rear, sufficiently close enough to allow for a defensive reaction to any aggressive act by the suspect.

Figure 8-8

Position, of course, is not the sole solution for officer safety. However, proper position, accompanied by select and constant suspect observation and perception, best affords our original goal of mutual safety for the officer and the citizen.

One Officer/Multiple Subjects

Again the officer should attempt to approach the closest and/or most threatening subject from the side and rear, position 2 or 10 o'clock.

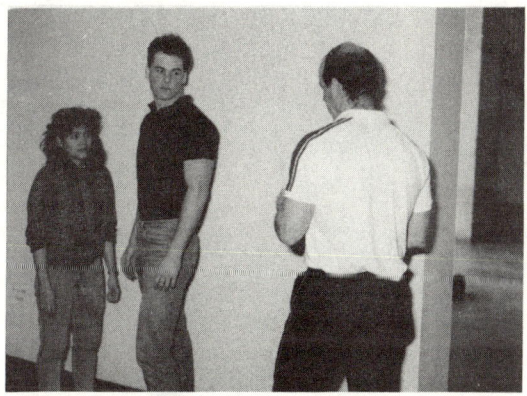

Figure 8-9

The officer should continue to utilize body language and tactical movements to shift this subject into a shielding position from the others present.

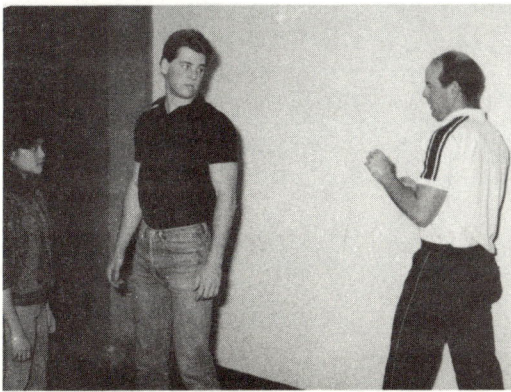

Figure 8-10

Chapter 9

PUNCHING, KICKING, BLOCKING, FALLING DRILLS

During your sequences of instruction in the various areas of defensive tactics, you will engage in numerous defensive drills. Drill structure allows for repetitive reinforcement of basic skills in a manner that facilitates a more efficient instructor critique of student achievement.

Punching Drill

The majority of physical encounters are defensive; however, it is felt that a portion of your training should include offensive strike capability. It must be emphasized that your employing agency may discourage or even disallow fist-striking techniques. However, by learning their basic elements perhaps you can enhance your knowledge and skills levels for defensive counters. In order to form a proper fist, fold all four of your fingers in as tightly as possible, and clamp your thumb down on top of them firmly. By folding each of your fingers in firmly, you provide greater protection to the finger's base and give increased tension to the wrist. (Figs. 9-1 through 9-4)

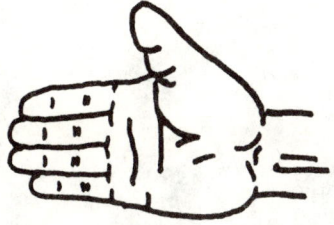

Figure 9-1

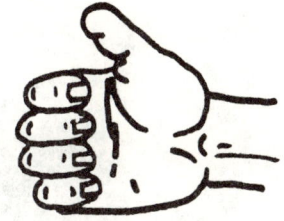

Figure 9-2

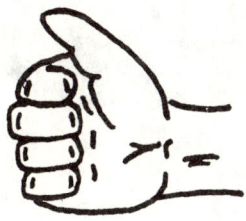

Figure 9-3

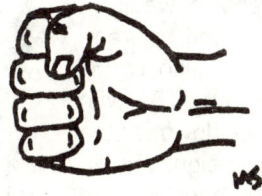

Figure 9-4

Since approximately 90% of the population is right-handed, we will define and detail the four basic punches—jab, cross, right and left hooks—on this basis. However, each technique can be easily adapted for the left-handed officer.

The following visual sequences will assist you in the accurate adherence of these techniques in regard to style as well as skill.

Left Jab

The left jab is thrown with a snapping action concluding with the arm fully extended. The jab should be executed with the palm down with the trajectory specific to the height of the aggressor. In unison with your fist, the left leg slides forward, followed by the right, trailing leg. Front and side views are shown correspondingly in Figs. 9-5 through 9-7.

Figure 9-5   Figure 9-6

Right Cross

Although the right hand can be the primary defensive weapon, parrying, catching and redirecting punches, etc.; it is also a powerful offensive weapon. As with the jab, the palm is down for the right cross. Your weight should shift to the right leg and allow the entire right side of your body to go into the punch. Instantly, return the right to its defensive position. Front and side views are shown correspondingly in Figure 9-8 through 9-10.

Figure 9-7   Figure 9-8

Figure 9-9   Figure 9-10

Left Hook

When you throw the left hook, the palm should face you. Your fist, elbow, and shoulder should be in line capitalizing on a snappy pivot to increase the power of the blow. The pivot should consist of a simultaneous twisting of the left hip and shoulder to the right. Front and side views are shown correspondingly in Figs. 9-11 through 9-13.

Figure 9-11

Figure 9-12

Figure 9-13

Figure 9-14

Right Hook

The motion is identical to that of the left hook. Again, the palm faces inward when you execute the hook with the right hand. Likewise, the pivot motion develops the punching power. Front and side views are shown correspondingly in Figs. 9-14 through 9-16.

Figure 9-15

Figure 9-16

PUNCH BLOCKING DRILLS

Slap Block

During any street encounter your hands should be held at a comfortable and acceptable height. The elbows should be held in close to the rib cage for mutual protection. (Fig. 9-17)

Figure 9-17

Therefore, in order to capitalize on this posture, the slap block will be the primary defensive counter to the straight jab due to its effectiveness and simplicity.

One maxim applicable to punch defenses is that generally the fastest, most effective method is to block with the hand closest to the aggressor's striking fist. For example, if you are in your alert stance with yur right leg to the rear, left leg to the front, etc., and the subject were to throw a right cross, then your left hand, due to its proximity would be the logical defensive source. The left hand then would redirect the force of the blow toward the center area of the officer's body.

Once the force of the blow has been redirected, reposition the left hand to its original placement so as to allow for additional blocking capability or the potential for an aggressive strike.

Obviously, this principle of blocking with the hand closest to the striking source is recommended for either side of the officer and regardless of his strong-hand/weak-hand profile.

The following visual sequences will correspond to the four most common punches, the left jab, right cross, left hook and right hook. Study the sequences closely since during the drill exercise you will assume both the offensive and defensive roles. (Figs. 9-18 through 9-21)

Figure 9-18
Perception of Left Jab

Figure 9-19
Right Palm Positioned
as Left Jab is Redirected

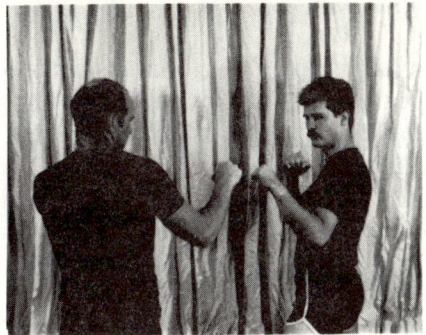
Figure 9-20
Perception of Right Cross

Figure 9-21
Left Palm Positioned as
Right Cross is Directed

Hook Block

The hook block is founded upon the premise of blocking with the hand and arm closest to the striking source. Again, from the alert stance, the officer simply moves his appropriate hand to the rear of his head while the elbow moves up the side of the face setting the forearm, elbow, and upper arm to absorb the force of the blow. Do not tilt your head down pulling your visual contact off the subject and increasing your risk of attack. Rather, bend your knees and crouch into a lower defensive posture, still maintaining your visual reference upon the source of threat. (Figs. 9-22 through 9-25)

LEFT HOOK BLOCK

Figure 9-22

Figure 9-23

RIGHT HOOK BLOCK

Figure 9-24

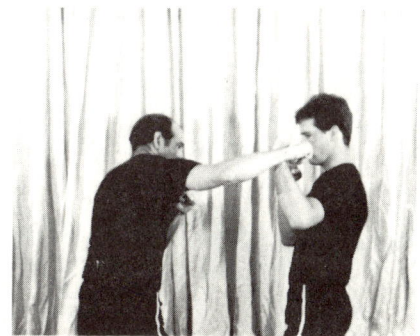
Figure 9-25

Kicking Drills

Again, it should be noted in preface that some police agencies prohibit the execution of foot blows by an officer. It is our feeling that kicking techniques are a must for self-protection and, in fact, superior to striking techniques with the hands for the following reasons:

1. The leg attack is longer and stronger than the arm. In most cases, an officer dramatically increases his power by calling on the kick rather than the punch. Kicks pack from three to five times as much destructive force as hand blows. Additionally, since the leg is much longer than the arm, the officer can maintain a greater distance from the subject and capitalize on maintenance of spatial safety.

2. Often, since the attack is visually directed toward the subject's lower torso, the target area is less elusive than the upper torso and head.

3. Lastly, kicking is an unexpected defense. The suspect ordinarily will, in actuality, overcompensate his upper torso defense and increase the vulnerability of a lower torso attack via foot-striking techniques.

As previously noted, regardless of the specific practicality of these techniques to you or your department, an increased awareness of foot techniques can prepare you for potential street encounters where the suspect's offense consists of a leg attack.

Now, as we begin to develop five basic kicking techniques applicable to an enforcement environment, we must review the principle of balance and reveal the factoral influence of speed. The increased power of a kick over a punch is lost if balance is not attained. If your kick lacks speed, it can be easily grabbed by the subject with devastating results. To develop speed, balance is essential; to develop balance, speed is essential. The two are completely inseparable. The two are completely unachievable if you are not willing to practice hour after hour; day after day; indeed month after month.

Five kicks have been selected to provide the greatest degree of efficiency and applicability to street situations. The five kicks include:

1. Front Snap Kick
2. Sike Kick
3. Front Angle
4. Roundhouse Kick
5. Rear Kick

Front Snap Kick

Raise the striking leg with the knee bent, keeping the lower part of the leg relaxed. The knee should actually become the sighting mechanism targeting the strike. Snap the lower leg out hard and fast. Once the strike has been attempted and the lower portion of the leg fully extended, snap the lower leg back into the original bent knee, tuck position before lowering it to the surface. Your body's balance can best be maintained by having a slight bend in the supporting leg. The front snap kick, like all the kicks you will practice, is to be targeted primarily at the shin and knee area of the subject, and only secondarily directed to the maximum height of the groin. (Figs. 9-26 through 9-29)

Figure 9-26

Figure 9-27

Figure 9-28

Figure 9-29

Side Kick

Raise the kicking leg so that the sole of the kicking foot is parallel to the inside of the knee of the supporting leg. Commence the kick in a straight line, striking the surface with the outer edge (knife edge) of the foot. Recover the kicking leg back into either a "cocked" position or in an alert stance. (Figs. 9-30 through 9-32)

Figure 9-30

Figure 9-31

Figure 9-32

Front Angle Kick

Here we make a modification in the target trajectory of the kicking leg. As with the side thrust kick, the sole of the kicking foot parallels the inner knee of the supporting leg. However, instead of a straight-line kick to the side, the kick is directed at approximately a 45° angle to the officer's side. One can enhance the power of the kick by the addition of a rolling hip motion of the kicking leg and a consecutive springing action of the supporting leg. This kick is excellent in warding off an attack by a "Third Party" who is attempting to thwart the arrest by the officer. (Figs. 9-33 through 9-35)

Figure 9-33

Figure 9-34

Figure 9-35

Roundhouse Kick

This kick is perhaps the most difficult due to the balance required for its execution; however, it is one of the most useful and powerful kicks one can master. Raise the kicking leg with a bend at the knee. The general striking area on the subject can be sighted via the knee's placement in the air. The knee now acts as a fulcrum, while the lower portion of the leg is made almost parallel with the floor. Now with a twisting role of the hips, swing the top of the striking foot, or ball of the foot where appropriate, into the target area. Upon the strike, return to the "cocked" position and then to the floor. (Figs. 9-36 through 9-40)

Figure 9-36

Figure 9-37

Figure 9-38

Figure 9-39

Figure 9-40

Rear Kick

Our previous kicks have been directed to counter attacks from the front or side of the officer. However, the rear kick is designed to negate an attack from behind by targeting an area between the solor plexus and knees of the attacker.

Initially, look over the shoulder of the designated kicking leg. Shift the weight to the supporting leg while drawing the kicking leg upward with the knee held at approximately hip level. The leg is then forcefully extended outward toward the target preferably striking with the broadest portion of the foot. It should be noted that a variation of the rear kick can easily be adapted to target the instep of the attacker via a simple stamping action. Rear kick techniques are illustrated in Figures 9-41 through 9-43

Figure 9-41

Figure 9-42

Figure 9-43

Blocks in Relation to Kicks

There are some key points that should be exposed in order to facilitate the successful mitigation of intrusions, specifically kicks. By virtue of the muscle mass of the limb, legs are more powerful than arms. Likewise, kicks are more powerful than punches. With this in mind, additional information must be pursued in regards to the officer's assessment of kicks.

One can assess the expertise of the subject's knowledge of martial arts or kicking skills, specifically, by the form of the kick—the chamber and the extension. As a general rule, kicks that initiate from the ground and have to ascend to come into contact with one, can be viewed as primitive. (Fig. 9-44 and 9-45) Although primitive, they should not be discounted for these kicks, in vital areas, are extremely injurious.

Figure 9-44

Figure 9-45

Refined kicks should follow a straight line. The axiom, "The shortest distance between two points is a straight line," is applicable to kicks and represents the ideal. This manifests itself in kicks that initiate at waist level (the chamber) and extend directly to the point of focus (the groin, for example.) (Fig. 9-46 and 9-47)

Figure 9-46

Figure 9-47

With this knowledge, the assessment of the expertise of the subject's kicking ability is a blatant asset to the officer for he or she can gauge his conduct accordingly. Furthermore, by having the ability to assess kicks, one has the opportunity to prevent injury to oneself. A kick that conforms to the ideal previously described of a refined quality should be viewed with considerable caution, for one strike to the face can shatter the jaw, teeth, and nose with minimal effort. In such situations, the officer's nightstick should be used as the first form of defense. (Fig. 9-48)

Figure 9-48

Because of the effectiveness of kicks, insured methods of blocking should be of considerable concern. Primarily, any intrusions into any quadrant of the body should be locked with the appropriate arm. Also, one should realize that the potential for success with one's strong arm against a kick is greater than that of one's weak arm.

Because of the strength and coordinational disparity, the strong arm is warranted and should be used if possible. (Fig. 9-49 and 9-50)

Figure 9-49   Figure 9-50

Kicks that are directed toward the officer's groin may be blocked with an appropriate arm block depending on the quadrant violated; however, the officer may discover that retracting the front leg to a "tuck" position may be quicker.

Although the kick has not been blocked per se, the withdrawing of the leg serves as protection to the groin as well as the abdominal area. It also serves to defuse the power of the kick. (Fig. 9-51 and 9-52)

Figure 9-51   Figure 9-52

Finally, an extremely effective means for the negation of a kick lies in the positioning of one's body. If the officer can detect bodily cues or body language that indicates that a subject is about to initiate a kick, the single most effective means of negation lies with advancing. While redirecting the threatening kick with the free hand or parying hand, the officer should charge the subject directing a single punch into an appropriate area (solar plexus). The charging into the subject has a variety of assets.

The initiation of kick necessitates the subject balancing on one foot. Since the balancing of the subject on one foot reduces his stability, he is extremely susceptible to being knocked down. As a rule, most inexperienced individuals, while initiating a kick, are extremely preoccupied with the kick and little else. Charging the subject will generally catch him unprepared and off guard. The punch that accompanies the charge serves as a deterrent to further confrontation. (Figs. 9-53 through 9-58)

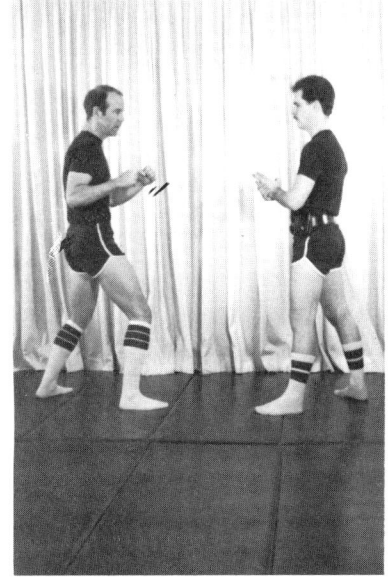

Figure 9-53   Figure 9-54   Figure 9-55

Figure 9-56　　　　　　Figure 9-57　　　　　　Figure 9-58

Additionally, it is advisable that immediately following the deterrent punch, the officer retreats and requests the subject to assume an appropriate controlled position. To follow the subject to the ground is inadvisable for a scuffle or a wrestling match is likely to ensue and the officer should avoid this at all costs. Also, following the subject to the ground may be interpreted as being precipitously aggressive in nature and may prompt defensive tactics on the subject's part that may prolong the confrontation.

Falling Drills

The ability to perform a fall in a controlled, safe manner will be essential during your defensive tactics instruction and invaluable in aggressive street encounters.

You will be given the opportunity to progress slowly through the developmental levels to minimize injury and maximize your skill acquisition.

Your falls will all begin low to the floor with a gradual increase in height, first starting in a static position and finally executed with motion.

Like all principles, in the most appropriate combative art, judo, falls follow the laws of nature concerning circular motion, roundness, and area absorption of shock. Here, the contact shock should be over as great an area as possible avoiding direct blows to the head or joints.

Your instruction in falling techniques will be confined to the following categories:

1. Back falls
   a. Supine Position
   b. Sitting Position
   c. Squatting Position
   d. Standing Position

2. Side falls
   a. Supine Position
   b. Sitting Position
   c. Squatting Position
   d. Standing Position

3. Front falls
   a. Kneeling Position
   b. Squatting Position
   c. Standing Position

4. Front Rolls
   a. Squatting Position
   b. Standing Position

BACK FALLS

Supine Position

Status: Arms should be crossd at the forearms and held across the chest. The hands should be opened with the fingers pointed forward. (See Fig. 9-59)

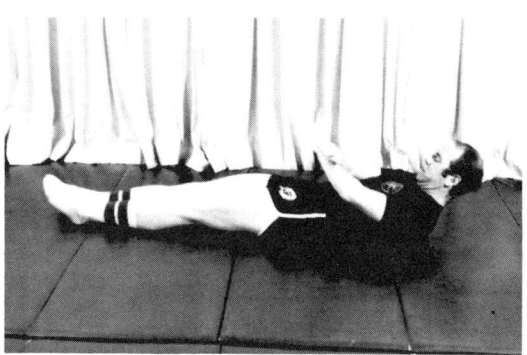

Figure 9-59

Motion: Slap the mat laterally at a 45° angle to the body, the palms down with both hands striking simultaneously. (See Fig. 9-60) Upon mat contact the elbows should be turned out with maximum force being absorbed by the palm and forearm. (See Fig. 9-61)

Figure 9-60

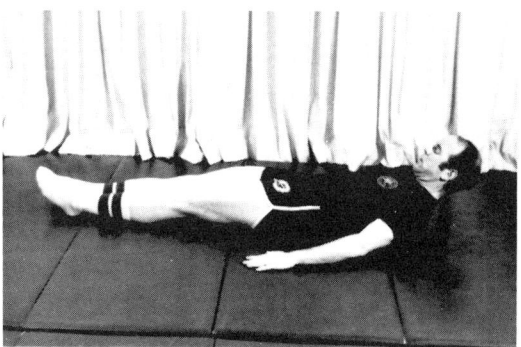

Figure 9-61

When on your back, always keep your head up, chin tucked forward, and eyes focusing on the naval region. Slap the mat with a springlike motion inductive of recoil action.

Sitting Position

Status: Sit with legs extended straight out in front, arms crossed at mid-chest height, palms down. (See Fig. 9-62)

Figure 9-62

Motion: Allow the upper torso to fall backward. (See Fig. 9-63)

Figure 9-63

The feet and legs should be held together and continue a path of natural upward movement corresponding to the action of the upper torso. As the upper back (scapula region) makes mat contact, the palms and forearms should yield a simultaneous slap. (See Fig. 9-64)

Figure 9-64

The recoil effect of the slap should return the body to the initial sitting position.

During this entire sequence, the entire body should conform to an imaginary circle without becoming tensed.

Squatting Position

Status: Assume a squatting position keeping the back relatively straight, arms crossed at mid-chest shoulder level, palms down. (See Fig. 9-65) The knees should be bent laterally out with the body weight balanced on the balls of the feet.

Figure 9-65

Motion: As the movement is initiated, an imaginary circle design for body conformation should be assumed. (See Fig. 9-66) Drop to the mat with the rounded area of the buttocks and lower back. The hand slap again should be proper and result in a reaction recovery to the starting position. (See Fig. 9-67) Recovery can be enhanced by quickly bringing the heels to the buttocks after the roll backward.

Figure 9-66

Figure 9-67

Standing Position

Status: From a natural standing posture, extend the arms in front at mid-chest level and move into a squatting position. (See Fig. 9-68) This movement from the standing to squatting position should be practiced so as to effect a smooth and fluid action.

Figure 9-68

Motion: Follow through with the backward motion contacting the mat with the "rounded body" design with first the buttocks, small of the back, and shoulder regions. (See Fig. 9-69)

Figure 9-69

The hand and arm slap should act to absorb the falling momentum and act as a spring to initiate a subsequent standing position if desired. (See Fig. 9-70 through Fig. 9-73)

Figure 9-70

Figure 9-71

Figure 9-72

Figure 9-73

SIDE FALL

Supine Position

Status: Assume a position on your back with both feet up in the air. The right hand is held up with the palm directed toward the knees. The left hand is lying on the left front of the hip. (See Fig. 9-74)

Figure 9-74

Motion: Pivot to the right side on your buttocks and initiate a hand slap with your right hand and arm. Simultaneously the outside of your right leg and foot should contact the mat, and the inside of the left foot should additionally absorb the shock. (See Fig. 9-75 and 9-76) Additionally, avoid contact with the small of the back and kidney region on the side fall.

Figure 9-75

Figure 9-76

Sitting Position

Status: Sit with both legs straight out in front, right hand stretched out at shoulder level, palm down; the left hand can remain lying on the left front hip region. (See Fig. 9-77)

Figure 9-77

Movement: Allow your body to fall to the right rear slapping the mat at a 45° angle from the body. Allow the feet to continue upward to full height of the momentum created. (See Fig. 9-78) Allow the downward force of the feet and legs to regain a controlled recovery to the original position. (See Fig. 9-79)

Figure 9-78

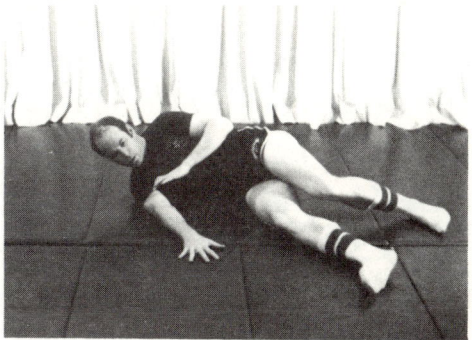

Figure 9-79

Squatting Position

Status: Assume a squatting position on the balls of your feet, with your knees bent laterally out. (See Fig. 9-80) During this phase, your back should be straight with your hands positioned as in the side fall from the sitting position.

Figure 9-80

Movement: The right foot is raised and moved over to a suspended position in front of the left foot. (See Fig. 9-81)

Figure 9-81

As the upper body falls to the right rear, the right hand and forearm slap. (See Fig. 9-82)

Figure 9-82

Standing Position

Status: From a normal standing position raise the left hand upward to head level with the palm in. (See Fig. 9-83) The right hand is positioned in a relaxed manner near your front, right hip.

Movement: Advance the left foot approximately 45° obliquely forward with the knee slightly bent. (See Fig. 9-84) As in the squatting position, move the left foot over to a position in front of the right foot. (See Fig. 9-85)

Figure 9-83　　　　Figure 9-84　　　　Figure 9-85

As the left knee increases its bend (See Fig. 9-86) until the upperbody falls to the left rear and side, the left hip region and left side region should sequentially land on the mat. (See Fig. 9-87) As the right hip makes contact with the mat, strongly strike the mat with your left palm and forearm. (See Fig. 9-88)

Figure 9-86   Figure 9-87   Figure 9-88

Again, the feet and legs roll upward upon landing. (See Fig. 9-89) The downward momentum of the legs after the fall was executed should be utilized to recover to the standing position if desired.

Figure 9-89

FRONT FALLS

Kneeling Positions

Status: The feet are extended behind the officer who has assumed a kneeling position. (See Fig. 9-90) The back is straight with the hands held in a front crossed position.

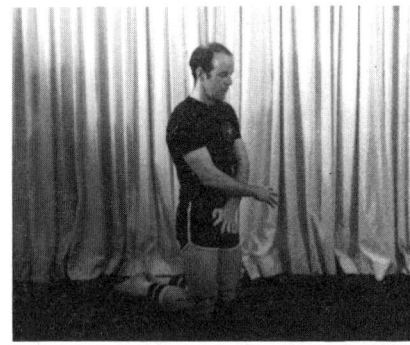

Figure 9-90

Movement: Allow the body to fall forward, preparing to absorb the shock with the hands (palms down) and forearms. (See Fig. 9-91) At the conclusion of the fall the hands should be forward of the shoulders with the elbows extended laterally at a 45° angle from the body. (See Fig. 9-92)

Figure 9-91

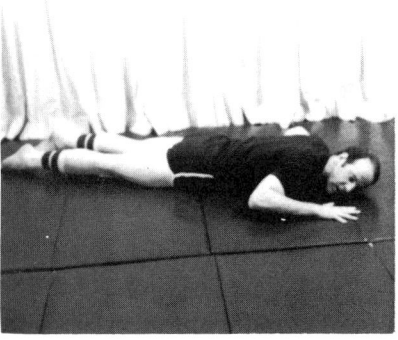

Figure 9-92

Squatting Position

Status: The knees are laterally out, weight maintained on the balls of the feet, with the back straight. (See Fig. 9-93)

Movement: Fall forward onto the mat with palms and forearms handling the shock absorption. (See Fig. 9-94 and 9-95)

Figure 9-93

Figure 9-94

Figure 9-95

Standing Position

Status: Assume a normal standing position. (See Fig. 9-96)

Figure 9-96

Movement: Slowly tilt forward and then fall toward the mat. (See Fig. 9-97)

Figure 9-97

The palms and forearms should simultaneously move upward to the upper torso region to accept the shock on impact. (See Fig. 9-98) Just before mat contact the face should be turned to the side. (See Fig. 9-99)

Figure 9-98

Figure 9-99

FRONT ROLL

Squatting Position

Status: Place your right hand palm on the floor approximately at a point directly in line below the chin. (See Fig. 9-100) The left hand should be placed between the legs, extended to the rear with the thumb pointed to the rear. (See Fig. 9-101)

Figure 9-100

Figure 9-101

Movement: As the left hand continues its rearward movement dip the left shoulder tucking your chin toward the upper chest. As your left roll commences, make chronological contact with your left wrist, your left elbow, left shoulder, left scapula, and concluding with the left hip. (See Fig. 9-102 and 9-103)

Figure 9-102

Figure 9-103

Simultaneously with the concluding contact with the left hip, and upon the feet contacting the surface, slap the surface with your right hand. (See Fig. 9-104 and 9-105)

Figure 9-104

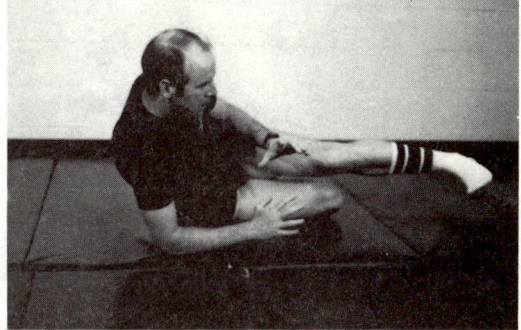
Figure 9-105

FRONT ROLL

Standing Position

Status: From a normal standing position bend forward at the waist while moving your hands into a circular configuration identical to their placement in the previous squatting position. By this action the body assumes a rounded configuration before commencing the left shoulder roll.

Movement: Upon contact of the palms on the floor the left elbow should bend rolling the body allowing for left shoulder contact with the surface. The rolling motion should continue its normal momentum with the leg motion allowing for a concluding standing position. (Fig. 9-106 through 9-114)

Figure 9-106

Figure 9-107

Figure 9-108

Figure 9-109

Figure 9-110

Figure 9-111

Figure 9-112

Figure 9-113

Figure 9-114

Figure 9-115

Chapter 10

PRINCIPLES OF ESCAPE

One of the most critical skills an officer can possess in his quest for personal protection is reaction time. And without question, the essence of escape is the officer's acquisition and maintenance of the highest degree of defensive, reactive sensitivity.

The officer must be able to react to the initial stimulus of attack which may manifest itself in blatant visual cues (i.e., winding up for a punch), or the obvious physical cue of a 220 pound drunk grabbing you from behind in a bear hug. And too, you must perceive the less blatant cues of an attitude shift, the covert move for superior positioning or the slightest touch commensurate with a rear assault.

Regardless of the clues or cues of aggressive behavior, to increase one's odds for the successful defense of an attack, the officer must be able to react to the initial stimulus in such a way to reduce the threat, the control or the injury. In other words, it is surely more difficult to break or escape from an already completed control hold or locking technique, than it is to counter the technique while it is still being effected and has not reached the stage of completion.

The point at issue is prioritized defensive concepts. The first concept is that it is easier to assume a defensive posture if one is *aware* that the initial attack is emminent, for the attack may in fact be prevented. Secondly, if the officer is not aware of an attack, he must react *beneficially* to the first stimulus of the attack to reduce potential injury. These two concepts are exemplified in a mutual situation where a punch is directed at an officer. If the officer is aware that a punch is emerging, he can successfully block it. If he is not aware of the punch until it is already underway, and is past a point of blocking capability, he can turn his face away from the fist reducing the impact and diffusing the power. The third and final concept directed toward attack situations which have passed the stages of awareness, and beneficially founded redirection or redistribution, is that of effecting an *escape* from the basis of collected concentration and effectively directed counter attack techniques.

Obviously, this chapter cannot list or depict every escape technique relative to every control hold an officer may confront. We have provided a few of the most common attacks, that because of their frequency, warrant your attention.

WRIST ESCAPE TECHNIQUES

Single Hand Grab: Crossgrip

In a situation in which a subject grips the officer in a single handed, cross grip, a variety of escapes are available. (See. Fig. 10-1)

Figure 10-1

Prior to the initiation of any wrist technique or escape, however, one concept is extremely effective in assisting any further action. If one's hand is limp and relaxed when a grip is applied, the grip will require a significant amount of strength to eliminate or manipulate for an escape because of the lack of resistance. The lack of resistance in the wrist necessitates additional strength to mitigate the grip. Consequently, upon the establishment of a grip to the wrist, the officer should immediately expand the diameter of his wrist by flexing his palm. The expansion of the fingers and the widening of the palm will increase the diameter of the wrist and consequently loosen the grip. (See Fig. 10-2)

Figure 10-2

Keeping with effective simplicity, the first technique initiates by the officer raising his hand to a position that is perpendicular to the subject's arm and directing the fingers at a 45° angle from his position. (See Fig. 10-3)

Figure 10-3

The officer should rotate the entire hand at the wrist until the fingers extend downward. This rotation should parallel the rotation of a hand that washes a window in a circular motion. This movement should be brisk, as the quickness of the technique is directly proportional to the effectiveness. (See Fig. 10-4)

Figure 10-4

Once the subject has relinquished his grip, the officer should step back and assume a defensive posture.

Single Hand Grab: Crossgrip

Another technique of considerable effectiveness employs the previous escape but elaborates upon it to transform it into a controlling technique. Upon the establishment of the cross grip by the subject, flex the hand to loosen the grip. (See Fig. 10-5 and 10-6)

Figure 10-5

Figure 10-6

At the same time, employ the free hand to brace the grip by pressing it against the hand that is gripping the officer's wrist. (See Fig. 10-7)

Figure 10-7

While holding the subject's grip tight against the wrist he holds, initiate the rotation utilized in the previous technique. (See Fig. 10-8 through 10-10)

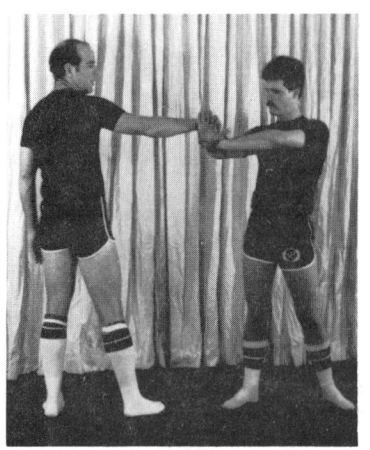  

Figure 10-8   Figure 10-9   Figure 10-10

When the rotation of the officer's wrist has reached shoulder height, grasp the wrist of the subject's arm that is applying the grip. Grasp the subject's wrist with the controlled hand. Insure that the edge of the officer's hand is as close to the subject's palm as possible. (See Fig. 10-11)

Figure 10-11

To achieve submission, tighten the grip on the subject's wrist and apply pressure downward. (See Fig. 10-12)

Figure 10-12

Double Hand Grip: Crossgrip or Sameside

Upon the establishment of a double hand crossgrip to the officer's arm, many viable techniques present themselves for utilization. (See Fig. 10-13)

Figure 10-13

Because of the level of strength that is applied with a two handed grip of any type, the flexing of the hand is required with a distracting technique such as a palm strike to the jaw of the subject. The flexing will loosen the grip, and the distractor will allow additional seconds to implement any type of technique. (See Fig. 10-14)

Figure 10-14

The escape technique requires the officer's elbow to remain stationary to eliminate a muscled escape in lieu of a leverage technique. The officer should raise his wrist directly toward his shoulder while leaning toward the subject. When the wrist has reached shoulder height, the officer should advance while raising the elbow to a position that is parallel to the ground. The grip should be virtually relinquished by the time the wrist reaches the shoulder, and the advancement and the raising of the elbow should act as an elbow strike to the sternum should the subject manage to maintain his grip. (See Fig. 10-15 through 10-17)

Figure 10-15   Figure 10-16   Figure 10-17

Double Hand Grip: Crossgrip or Sameside

The next technique employs the same principles as the first; however, it allows for muscular assistance. Upon the establishment of the double handed crossgrip, the officer should flex the wrist to loosen the grip. (See Fig. 10-18 and 10-19)

Figure 10-18   Figure 10-19

The officer should then reach between the subject's arms and grasp his own controlled hand in a handshake type grip. (See Fig. 10-20 through 10-22)

Figure 10-20 Figure 10-21 Figure 10-22

The officer's elbow should remain in basically the same position while employing muscular strength as well as leverage to pull the wrist toward the shoulder and free the controlled arm. (See Fig. 10-23 and 10-24)

Figure 10-23 Figure 10-24

Single Hand Grip: Sameside

Upon the establishment of a "sameside" grip, the officer should immediately flex his wrist to loosen the subject's grip. (See Fig. 10-25 and 10-26)

Figure 10-25

Figure 10-26

The officer should rotate his wrist and hand to a position that has the hand parallel to the floor with the palm facing up. (See Fig. 10-27)

Figure 10-27

Upon the position of the officer's hand, he should pull his entire arm and hand toward his shoulder which should split the subject's grip between the fingers and thumb. (See Fig. 10-28)

Figure 10-28

The sequencing of each of these maneuvers should be done as precisely as possible, yet as quickly as possible. The quicker and smoother the technique is applied, the more inclined the technique will be successful.

Single Hand Grip: Sameside

Upon the establishment of the sameside grip, the officer should flex his wrist. (See Fig. 10-29 and 10-30)

Figure 10-29

Figure 10-30

The officer should then raise his hand to a position that bisects the subject's arm with his palm. His thumb should be on one side of the arm with his fingers on the other, and with the controlled hand, the officer should grasp the subject's arm. This should eliminate the grip. (See Fig. 10-31 and 10-32)

Figure 10-31

Figure 10-32

With the free hand, the officer should grasp the subject's hand with his palm across the backside of his hand and his fingers along the edge of the subject's palm. The officer's thumb should be around the thumb side of the subject's hand with his thumb above that of the subjects. (See Fig. 10-33)

Figure 10-33

The officer should rotate the subject's hand so that the edge of his hand is pointing directly up. The hand that was originally grasped by the subject should be placed so that the webbing between the thumb and index finger is directly beneath the subject's thumb and a grip is assumed from that placement. (See Fig. 10-34)

Figure 10-34

Submission is achieved by simultaneously pulling down on the wrist and pushing the edge of the subject's hand directly toward his face. (See Fig. 10-34 and 10-36)

Figure 10-35

Figure 10-36

Rear Double Wrist Grip

Immediately upon the establishment of the grip, the officer should flex his wrists. (See Fig. 10-37 through 10-39)

Figure 10-37　　　　　　　　Figure 10-38　　　　　　　　Figure 10-39

Once the wrists are flexed, the officer should extend his arms outward at approximately 45 degree angles from his body. (See Fig. 10-40 and 10-41)

Figure 10-40　　　　　　　　　　　　　　　　Figure 10-41

While continuing to raise the arms, the officer should take a step backwards and away at a 45 degree angle from his original position. (See Fig. 10-42 and 10-43)

Figure 10-42

Figure 10-43

At this point, the officer should let one arm drop while raising the other arm and rotating. (See Fig. 10-44 and 10-45)

Figure 10-44

Figure 10-45

The officer should now be at the backside of the subject. The arm that was low should continue rotating as should the top arm. (See Fig. 10-46 and 10-47)

Figure 10-46

Figure 10-47

The arm that was low should now be raised to assume a position that is directly behind the elbow of the subject's outside arm. The edge of the officer's hand should be perpendicular to the subject's arm. (See Fig. 10-48)

Figure 10-48

With the hand that is behind the subject's elbow, push the arm away and raise the other wrist to eliminate the grip. Upon establishing a defensive posture, this concludes the technique. (See Fig. 10-49 through 10-51)

Figure 10-49

Figure 10-50

Figure 10-51

Rear Double Wrist Grab

Upon the establishment of the rear double wrist grab, the officer should flex his wrists as mentioned previously. (See Fig. 10-52 and 10-53)

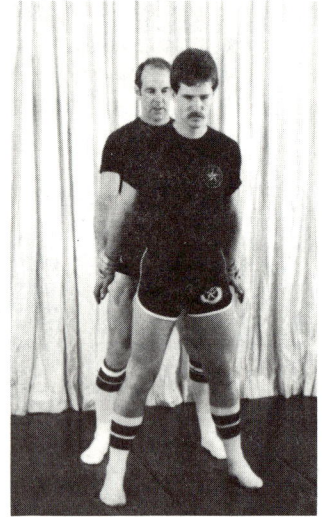
Figure 10-52

Figure 10-53

The officer should turn toward the subject letting one arm trail behind him, while his other arm assumes a position that is bent at the elbow and parallel to the ground. The arm that is parallel to the ground should strike the subject in the underside of his elbow. The rotation of the officer toward the subject, in combination with the arm position, should create an elbow strike to the subject's lower bicep area. (See Fig. 10-54 through 10-56)

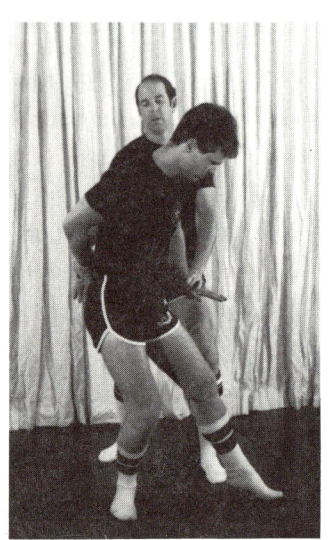
Figure 10-54

Figure 10-55

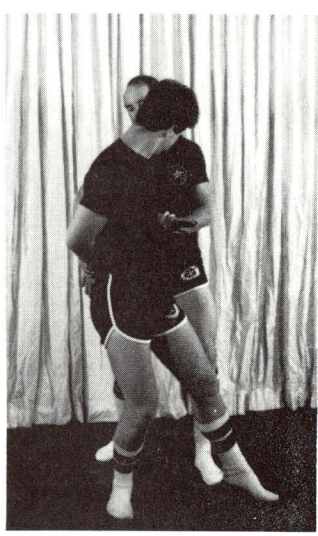
Figure 10-56

If, at this point, the subject does not relinquish his grips on the wrist, the officer must assume (as with any rear attack) that the subject is attempting to disarm him of his service revolver. The officer should act accordingly with defensive techniques such as a strike to the groin followed by a backfist to the bridge of the nose. (See Fig. 10-57 through 10-59)

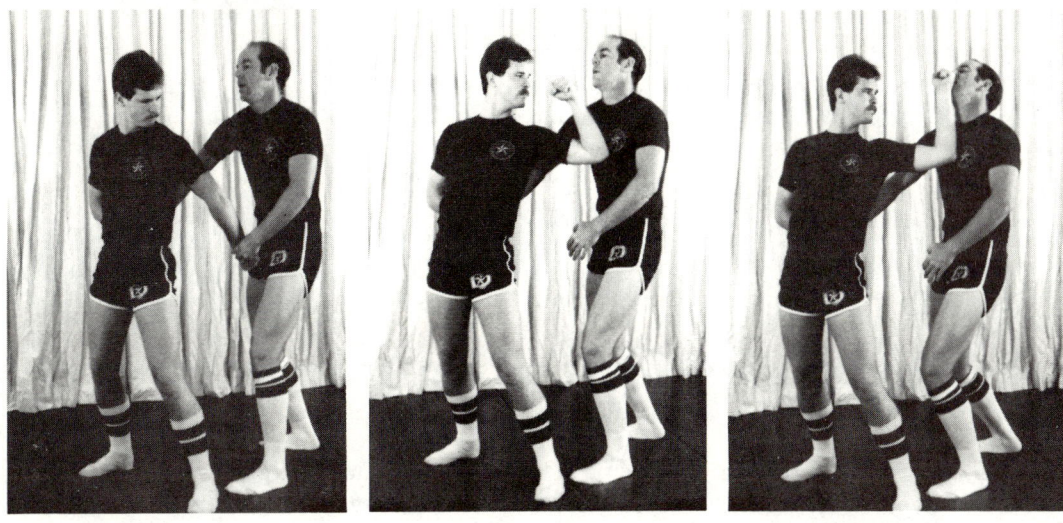

Figure 10-57　　　　Figure 10-58　　　　Figure 10-59

Rear Double Wrist Grab

Once the subject has established a double wrist grab from the rear, the officer should flex his wrists. (See Fig. 10-60 and 10-61)

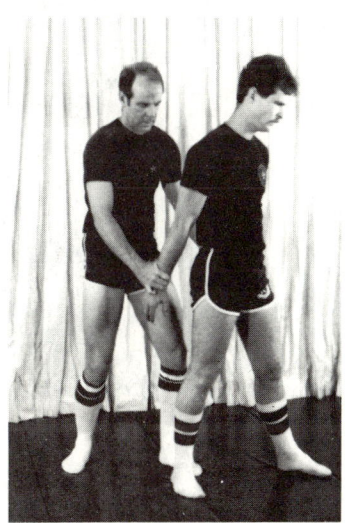 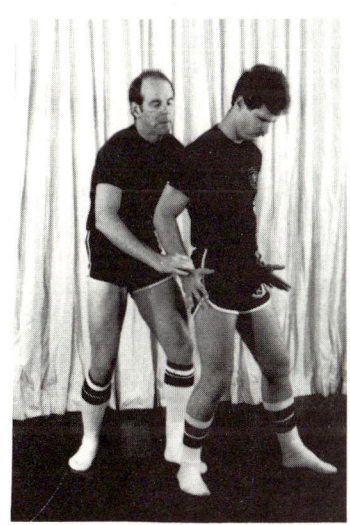

Figure 10-60　　　　　　　　　　Figure 10-61

The officer should immediately lower his center of gravity by bending his knees and should bend his arms upwards at the elbow. The arms should move toward the shoulders which will break the subject's grips. (See Fig. 10-62)

Figure 10-62

Now that the grips on the officer's wrists have been eliminated, a variety of defensive techniques may be employed to complete the escape such as an elbow strike to the ribs. (See Fig. 10-63 through 10-65)

Figure 10-63   Figure 10-64   Figure 10-65

GENERAL ESCAPE TECHNIQUES

Full Nelson: Prevention

If an individual approaches the officer from the rear, attempting to implement a full nelson, the single most effective deterrent is to lower the center of gravity by bending the knees. Simultaneously, the officer should pull his elbows into his ribs to prevent the subject from getting his arms under the officer's armpits. (See Fig. 10-66 and 10-67)

Figure 10-66

Figure 10-67

If the subject manages to implement the full nelson, a variety of defensive techniques are applicable. The officer may reach above his head to that of the subject's and grasp an ear which may be used to pull the subject off of his back. (See Fig. 10-68 and 10-69)

Figure 10-68

Figure 10-69

The officer may attempt a variety of kicks to the subject as a kick to the knee or groin. (See Fig. 10-70)

Figure 10-70

The officer may attempt a foot stomp to eliminate the grip. (See Fig. 10-71 and 10-72)

Figure 10-71   Figure 10-72

The officer may also employ a tripping technique from this position by lowering the center of gravity and stepping back and away. (See Fig. 10-73 and 10-74)

Figure 10-73   Figure 10-74

Once the officer has stepped behind the subject, he should place the leg that stepped back, between the subject's legs and should be in a fully crouched position. (See Fig. 10-75)

Figure 10-75

From this position, the officer should push forward with his hips and backward with his upper torso to trip the subject across his hip. (See Fig. 10-76)

Figure 10-76

Rear Choke Hold

If a subject should place the officer in a rear standing choke hold, the first priority of the officer should be to secure an airway. If the officer's respiration is cut off, defensive techniques have little effectiveness. This may be done by stabilizing the choking elbow and by positioning the chin below the inside of the elbow. (See Fig. 10-77 through 10-79)

Figure 10-77

Figure 10-78

Figure 10-79

From this position, the officer may employ a variety of techniques such as an elbow strike to the ribs. (See Fig. 10-80)

Figure 10-80

A groin strike may be employed. (See Fig. 10-81)

Figure 10-81

The officer may wish to remove the hold in such a fashion as to control the subject once it is removed. This may be done by grasping the wrist of the choking arm with his free hand. (See Fig. 10-82)

Figure 10-82

Once this has been done, the officer should slide to the side of the subject while pushing forward on the elbow and pulling on the wrist. This will free the officer. (See Fig. 10-83 and 10-84).

Figure 10-83

Figure 10-84

The officer may choose to throw the subject off of his back depending on the muscular capacity of the subject. This may be done by lowering his center of gravity and by grasping the choking arm. (See Fig. 10-85 and 10-86)

Figure 10-85

Figure 10-86

Once this has been completed, the officer should simultaneously bend at the waist and point his shoulder to a position directly in front of him on the ground.

Standard Head Lock

If the officer finds himself in a standard headlock position, there are many effective techniques to combat this. The first priority is the airway maintenance, however. (See Fig. 10-87)

Figure 10-87

The officer may attack the groin area since his hands are free. (See Fig. 10-88)

Figure 10-88

The officer may reach over the subject's head and grasp his nose for leverage in pulling him backwards. (See Fig. 10-89 through 10-91)

Figure 10-89   Figure 10-90   Figure 10-91

The officer may also employ an escape that first allows the officer to secure the subject's capturing arm at the wrist. Upon this stabilization, the subject's affected arm is pushed forward of the head. The officer concludes to the rear of the subject able to force him to the ground. (See Fig. 92 through 95)

Figure 10-92

Figure 10-93

Figure 10-94

Figure 10-95

Rear Bear Hug: Underarms

If the officer should find himself in a position in which a bear hug is implemented by a subject, the most effective techniques are the simplest. (See Fig. 10-96)

Figure 10-96

109

The hug may be eliminated by selecting a finger of the subject's and prying it loose to mitigate the grip. (See Fig. 10-97 through 10-99)

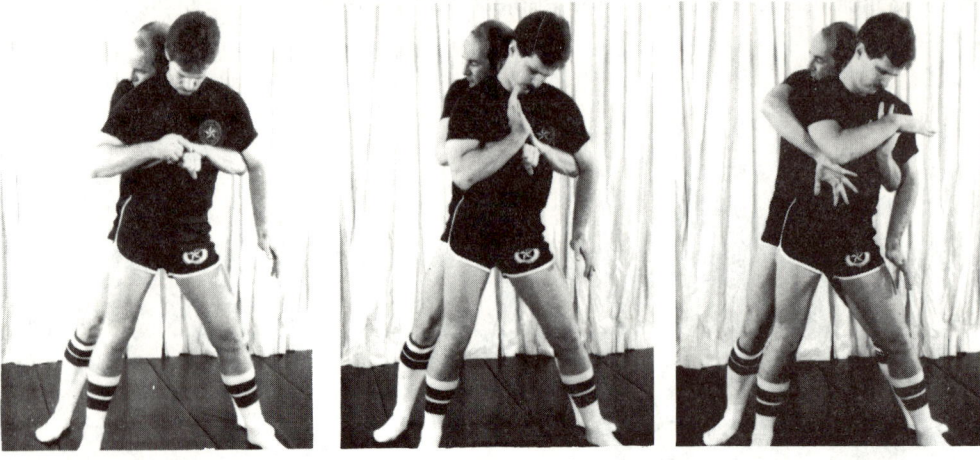

Figure 10-97   Figure 10-98   Figure 10-99

Once the grip is eliminated, the officer should turn away from the subject and attempt to get behind him. This may be done by using the officer's arm as a wedge for rotation. (See Fig. 10-100)

Figure 10-100

Rear Bear Hug: Overarms

The simplest and most effective technique to combat this hold is to attack the subject's groin with the free hands. (See Fig. 10-101 and Fig. 10-102)

Figure 10-101   Figure 10-102

Front Strangle Hold (1)

If the subject establishes a front strangle hold, the first priority is again clearing the airway. The escape technique may be initiated by raising an arm above the subject's and by stepping backwards. (See Fig. 10-103 and 10-104)

Figure 10-103

Figure 10-104

Once the officer has stepped back, he should rotate his shoulders so that he is perpendicular to the subject and his raised arm is near his opposite shoulder. (See Fig. 10-105)

Figure 10-105

This movement should break the strangle hold and the officer should initiate a deterrent strike such as an elbow strike to the face of the subject. (See Fig. and 10-107)

Figure 10-106

Figure 10-107

Front Strangle Hold (2)

The officer may also eliminate the grip by simply placing one finger in the jugular notch at the base of the subject's neck and pushing the subject away. (See Fig. 10-108 through 10-110)

Figure 10-108

Figure 10-109

Figure 10-110

Rear Strangle Hold (1)

This technique may be eliminated by a swift kick to the knee of the subject. (See Fig. 10-111 through 10-114)

Figure 10-111

Figure 10-112

Figure 10-113

Figure 10-114

Rear Strangle Hold (2)

The officer may also eliminate the hold by stepping away from the subject and assuming a defensive posture. (See Fig. 10-115 through 10-117)

Figure 10-115

Figure 10-116

figure 10-117

Chapter 11

STANDARD SEARCH TECHNIQUE

Needless to say, it is of paramount importance that the search of a suspect is efficient and totally effective. It is with this dual goal of efficiency and effectiveness that a standardized technique has been provided. It is, of course, impossible to develop one procedure that will conform to all situations of searching; however, this technique is the basis for other searching modifications to be presented later.

Spread-Eagle Wall Search

The most "time-tested" technique calls upon the suspect to initially face a non-reflective, supportive wall. The subject is told to place both hands, palms flat, as high and wide on the wall surface as possible. It is important to remember that the height of the hands is proportionately more debilitating upon the suspect than the hand width placement.

Simultaneous with the hand placement, the subject should be instructed to move his feet away from the wall with both feet remaining parallel to the wall and spread as wide as possible. (See Fig. 11-1)

Figure 11-1

The officer's position during the placement process should be directly to the rear of the suspect, adjusting his stance of safety as the subject moves, so as to maintain defensive distance. From this position, the officer constantly holding his alert stance, can best avoid side punches or kicks and visually detect most effectively any aggressive movement at its genesis. Your visual target should remain in the mid-shoulder blade area throughout the entire search sequence, maximizing your peripheral vision to key attention on all four limbs.

Once the subject has followed your instructions and has assumed a proper search stance, you are ready to commence the search. Emphasis must be added at this time to caution that the officer *must not proceed further unless he truly has the subject in a position of control.* Cues of this control can include foot slippage on the part of the suspect, quivering of the upper leg muscles, obvious induced impairment of balance, etc.

SEARCH SEQUENCE

Quadrant #1—Upper Left Torso

Under normal conditions, the actual commencement of the search begins on the subject's left or presumed weak side. This weak side initial contact allows for conclusion on the suspect's right side and; therefore, properly placed for eventual handcuffing. (See Fig. 11-2)

Figure 11-2

The approach consists of the officer placing his left foot over the subject's left foot so that his left heel is against the subject's left instep. The officer's right hand should simultaneously securely grasp the rear midbelt area of the suspect to control his center of gravity and additionally create psychological as well as physical control. (See Fig. 11-3) During this preliminary foot and hand placement, the officer's left hand and arm should be held high to ward off the left elbow or back-fist attack. (See Fig. 11-4)

Figure 11-3

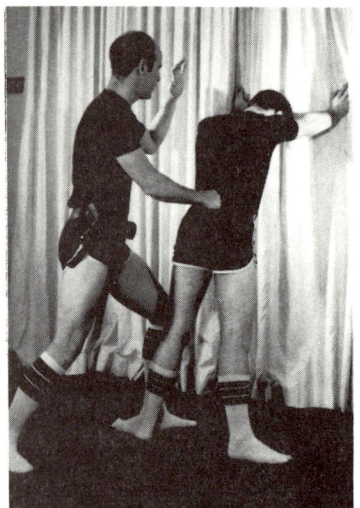

Figure 11-4

From this position, the actual search procedure can begin. Remember, searching is a phenomenon of touch; sight can act as a barrier to your effectiveness. Touch the area by crushing the surface, feel for the unnatural, squeeze the material right down to, and inclusive of, his skin. Your street search can, and should be as effective as a strip search with the exclusion of the body cavities.

First, lift the subject's left hand off the wall and check it for weapons or contraband. (See Fig. 11-5) Crush the clothing along, under and over the arm onto the left chest and then left back regions. (See Fig. 11-6)

Figure 11-5

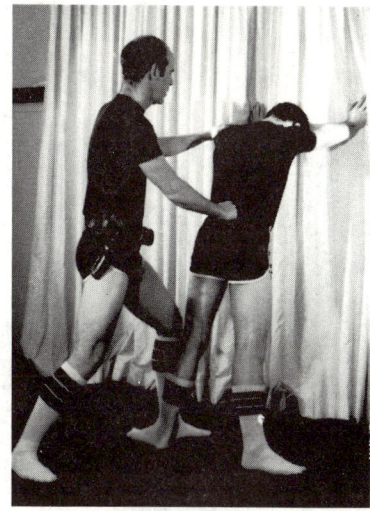

Figure 11-6

Check the neck area and verify the threat potential of articles including chains, necklaces, or suspended objects. (See Fig. 11-7) If a threat exists, remove the article. Move up the neck and check the ear area and conclude by raking the fingers through the hair. (See Fig. 11-8)

Figure 11-7

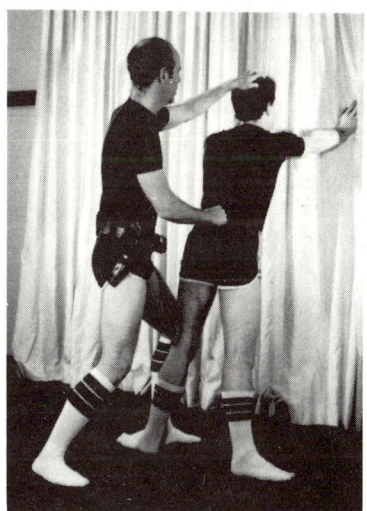

Figure 11-8

The main thrust of our searching approach is that it is a system by which safety is most successfully maintained. The systematic search of the upper left torso should extend beyond the parameters of first quadrant to assure double coverage of critical areas.

Quadrant #1 is concluded with increased emphasis toward the high-risk belt regions; the obvious catchall for weapons, etc. The waist presents layers of circular search direction. Use your fingertips to scan the skin surface around the waist. (See Fig. 11-9) Next, peruse the inner, then outer, band of the underwear; onto the inner, then outer, band of the trousers. (See Fig. 11-10) Conclude with a dual surface check of the subject's belt.

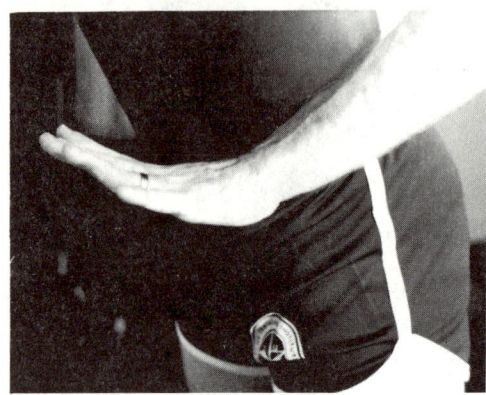

Figure 11-9

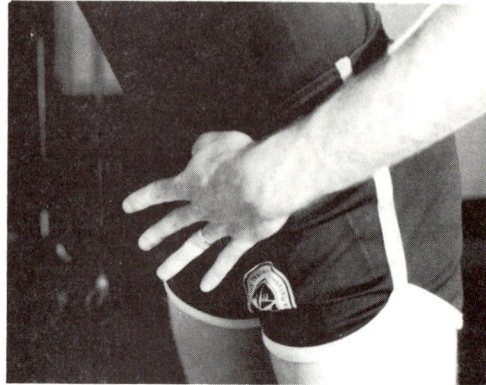

Figure 11-10

Quadrant #2—Lower Left Torso

Still controlling the suspect at the midbelt with your strong hand, pivot your feet smoothly (See Fig. 11-11) into the following position: your right foot moves in a sliding action so that your right instep is now placed next to the suspect's left heel (See Fig. 11-12), while your right knee is situated directly against the back of the subject's left knee. (See Fig. 11-13) Naturally, to assure a resumption of balance, move your left leg to the rear, imitating an "alert" style stance as the searching effort continues. (See Fig. 11-14)

Figure 11-11

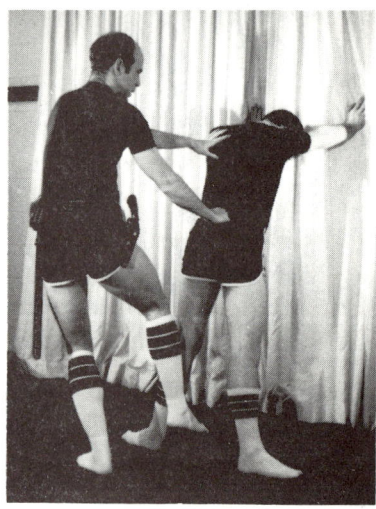

Figure 11-12

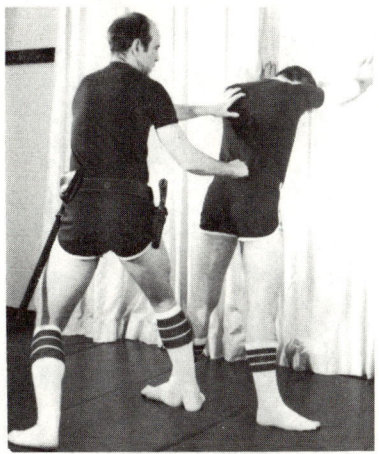

Figure 11-13

Figure 11-14

Initiate the lower torso search by a full resumption of the intensive layered inspection of the midsection. (See Fig. 11-15)

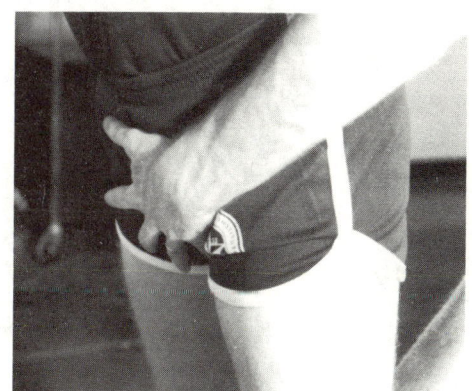

Figure 11-15

Descend with the inspection effort over the hips, buttocks and the groin. (See Fig. 11-16)

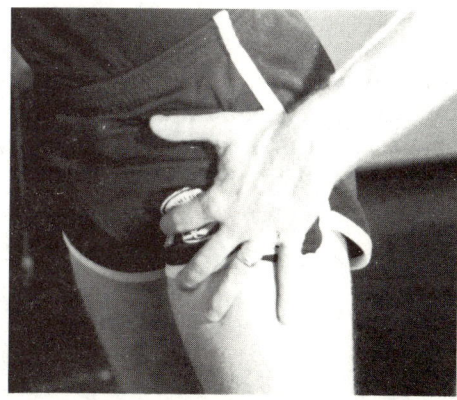

Figure 11-16

Concentrate on keeping your head up with your targeted vision center in the upper center back so the totality of the body is observed. (See Fig. 11-17)

Figure 11-17

Conclude your search of the lower torso with a systematic coverage of the entire leg down to the ankle. (See Fig. 11-18)

Figure 11-18

Although the suspect could easily secrete items in or under his shoes due to the tentative positioning, this examination had best be left until later when the shoes can safely be removed.

We suggest that if during this searching operation an item is found that must be removed from the subject, it should be repositioned, if safety allows, on a paralleling basis on your person. For instance, you discover items in the left front trouser pocket that may be supportive evidence indicative of theft; place the items into your left front trouser pocket. This procedure is much more secure than throwing the items on the ground, etc., and can aid in memory stabilization when that location of detection becomes a critical entry in your offense report.

Quadrant #3—Upper Right Torso

After searching fully the area of quadrant #2, step back to the rear of the suspect into your normal alert stance. (See Fig. 11-19)

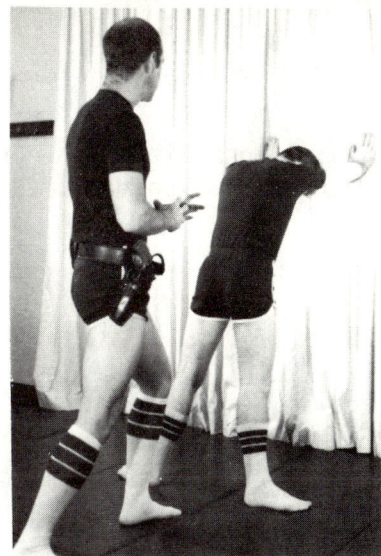

Figure 11-19

Take this opportunity to visually recheck the subject's position to assure that in the process of the left side search, he has not moved into a more secure and, therefore, for you a less safe status.

Now that you have examined the subject wall position and found it to your satisfaction, you can now proceed to search the right side of the suspect. Note this is presumed to be the normal "strong side" and that which presents the maximum threat.

The approach is similar to that used on the two previous quadrants, except opposite hands and feet are utilized. This time the officer's right foot is placed over the subject's right foot so that the officer's right heel is against the suspect's right instep. (See Fig. 11-20)

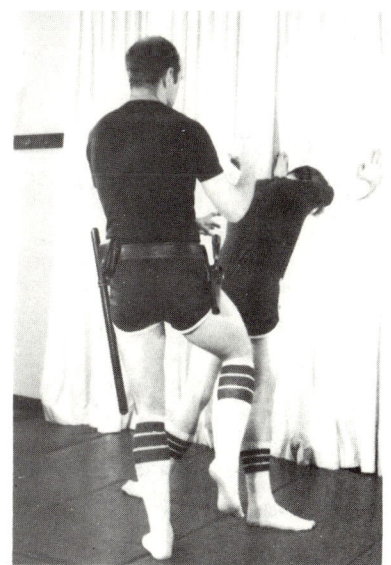

Figure 11-20

The officer's right hand and arm are upheld to ward off an elbow and backfist punch. (See Fig. 11-21)

Figure 11-21

The left hand is securely placed in the midcenter belt area to control the suspect physically and psychologically during the searching of both quadrant #3 and #4. (See Fig. 11-22)

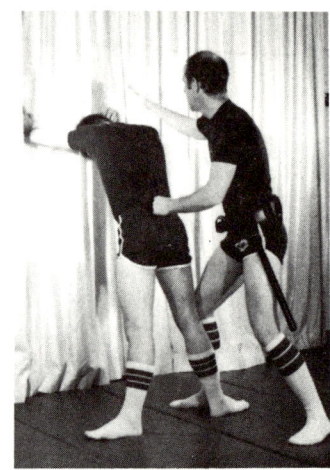

Figure 11-22

The systematic search now commences on the subject's upper torso, right side to the same degree and depth as detailed and depicted earlier for the left side of quadrant #1. (See Fig. 11-23 through 11-28)

Figure 11-23   Figure 11-24   Figure 11-25

 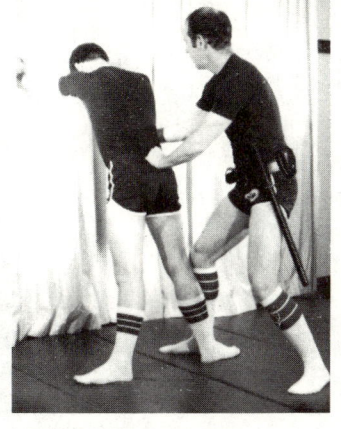 

Figure 11-26 Figure 11-27 Figure 11-28

Quadrant #4—Lower Right Torso

Again, a paralleling but opposite foot positioning occurs to search the lower torso, right side. The left foot is moved into a position so that the left instep is placed against the right heel. (See Fig. 11-29)

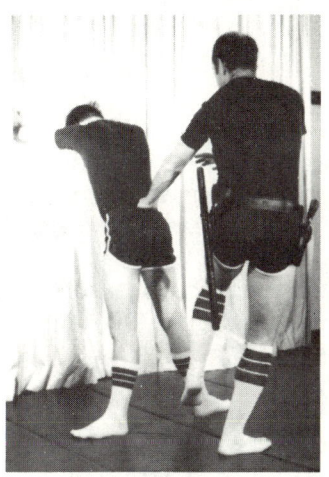

Figure 11-29

The left knee is to the rear and applies pressure on the subject's right knee. (See Fig. 11-30)

Figure 11-30

123

The right leg is positioned to the rear in the assumption of the standard alert stance. (See Fig. 11-31)

Figure 11-31

Essential to the officer's safety is the maintenance of balance throughout the search but especially in regard to the lower torso. Balance can best be achieved by bending your knees, not your back, keeping your head up and vision keyed on the targeting sight on the upper back.

Once again, the lower torso is searched by a systematic mode of coverage, as depicted earlier for the left side. (See Fig. 11-32 through 11-36)

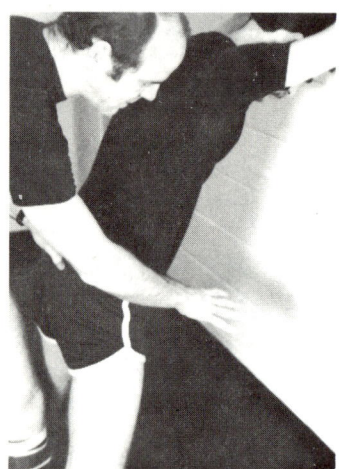  

Figure 11-32   Figure 11-33   Figure 11-34

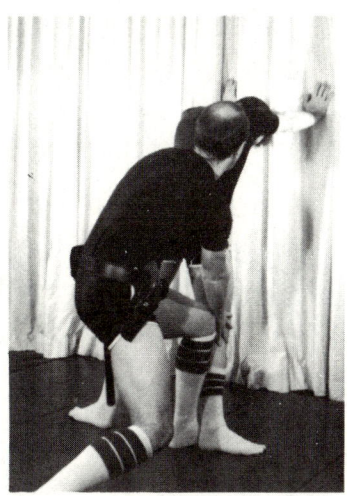
Figure 11-35

Figure 11-36

The officer assumes the alert stance to the rear of the searched subject in anticipation of eventual handcuffing options. (See Fig. 11-37)

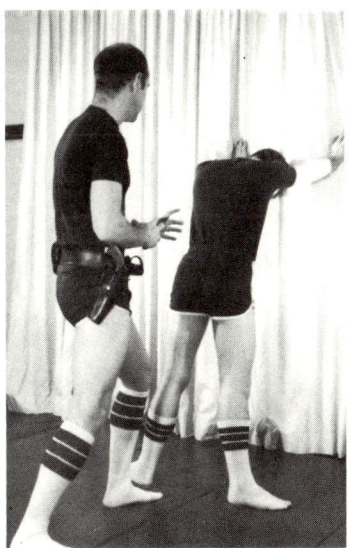
Figure 11-37

Chapter 12

NOMENCLATURE AND BASIC PROCEDURES OF HANDCUFFING

Handcuffs should be a tool well understood by all officers since only with full knowledge of their idiosyncracies can we maximize their effectiveness.

Quality handcuffs follow the same general design consisting of (See Fig. 12-1):

1. Double bar—The dual arch-like extension off the main frame of the handcuff.

2. Single bar—Lower movable secondary structure that flows through the double strand to accomplish the wrist restraint. Outermost portion consists of jaw or step-like cuts that are received by a ratchet within the body of the main frame.

3. Two link chain—Normally a dual or tri-link portion which maintains standard handcuff separation.

4. Lock entry mechanism—The receiving port for the nonhandcuff key which allows possible disengaement of double-lock status and actual single strand release.

5. Double lock hole and activator—This round intended part when pushed inward activates double lock capability inside ratchet body, increasing pick resistance and locking travel of single strand through double strand.

6. Standard handcuff key with spike end—This metal key has a simple cylinder and block end for alignment in lock mechanism for release potential. On the opposite end, a quarter-inch spike is designed to initiate the double-lock mode.

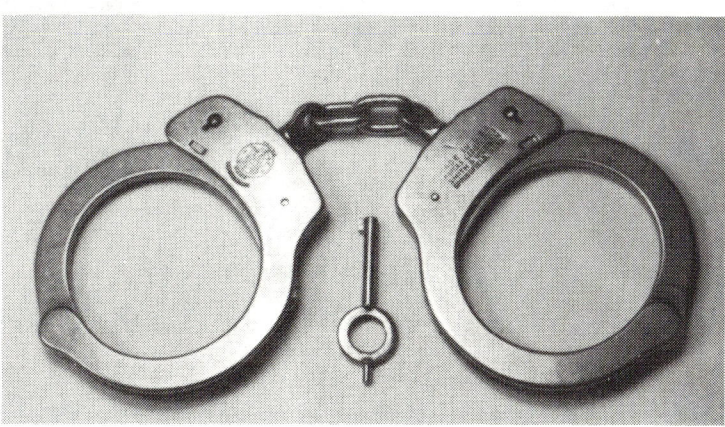

Figure 12-1

Additionally, you must understand that handcuffs are temporary restraining devices; they do not guarantee total control of any individual and; therefore, in no way should reduce your conscious perception of the threats of physical harm toward you by the suspect. Further, like your service revolver, they must be kept clean and constantly checked for proper operation.

Carrying and Positioning of Handcuffs

It is suggested that you carry your handcuffs in a covered cuff case to protect them from foreign elements and prevent loss. If you wear a chain-down cuff case, the handcuffs should be placed lock-entry mechanism to lock-entry mechanism, chain-down with single strand forward. (See Fig. 12-2)

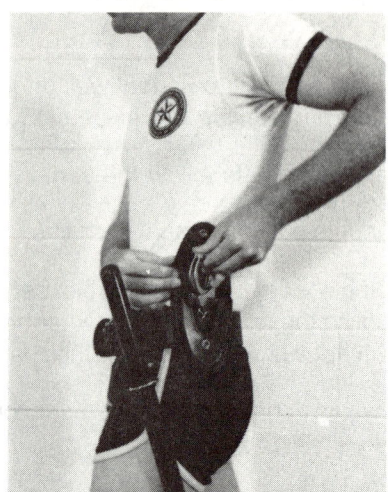

Figure 12-2

If you use a chain-up cuff case, carry the handcuffs lock-entry mechanism to lock-entry mechanism, chain-up with the double strand forward. (See Fig. 12-3)

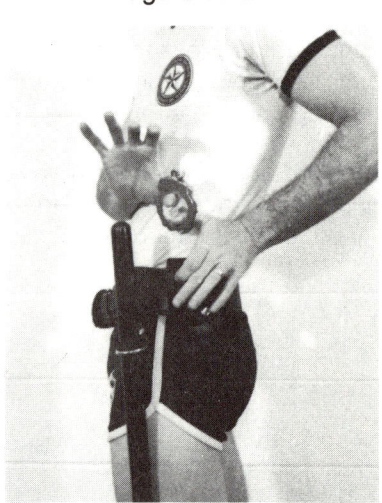

Figure 12-3

The handcuff holder case should generally be worn on the officer's left side in the approximate frontal hip area. (See Fig. 12-4)

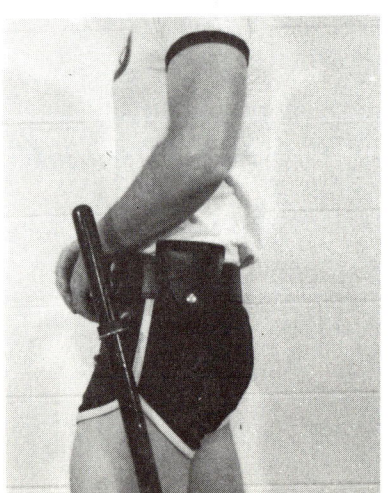

Figure 12-4

Standard Handcuffing Procedures

Generally, the best procedure is to search and then handcuff. Those exceptions to this rule will be noted later. Let us proceed through the handcuffing procedure under the assumption that a quality search has already been conducted. We have concluded the search and are now standing in our alert stance after just completing the search of the subject, upper and lower right torso in a normal stance or preferably with the baton in an on-guard position. (See Fig. 12-5)

Figure 12-5

Since 90% of the population are strong-hand-right, it is best to secure and handcuff the right wrist first.

To begin the handcuffing sequence, ask the subject to remove his right hand from the wall surface and slowly move it to the small of his back. (See Fig. 12-6 and 12-7)

Figure 12-6

Figure 12-7

Do not reach for the hand, stay in your alert stance with your elbows in and hands held high throughout the entire request sequence and compliance. Once the hand arrives at the noted location in the small of his back, reach over the suspect's right hand, grasping the index and middle fingers. (See Fig. 12-8)

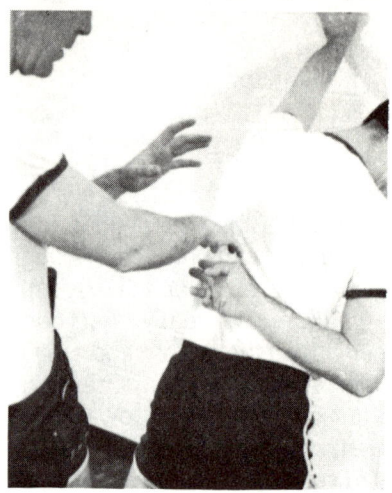

Figure 12-8

Pull up and twist these fingers gently toward the subject's back. In this state of exstension, the subject actually determines the degree of tension and correlated pain via his resistance. (See Fig. 12-9)

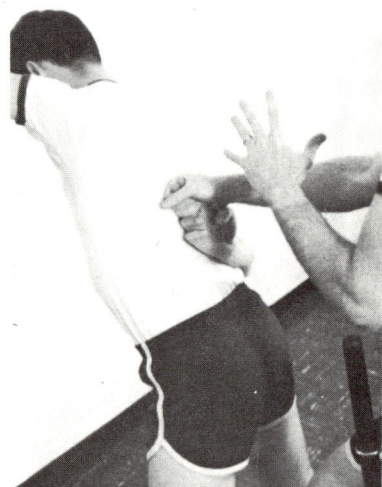

Figure 12-9

Once you feel a secure hold on the fingers and have succeeded in aligning the right wrist in a handcuff acceptance station, reach for your handcuffs on the left side of your body with your left hand. Grasp the outermost handcuff by placing your thumb on the double strand and your index finger around, and supportive of, the lock opening housing. (See Fig. 12-10)

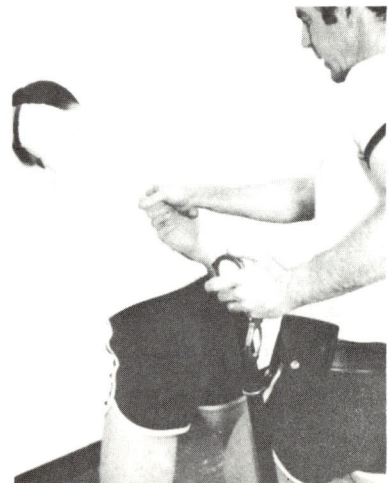

Figure 12-10

Move the handcuff to a point above the subject's right wrist and push, do not strike, the single strand against the wrist. (See Fig. 12-11)

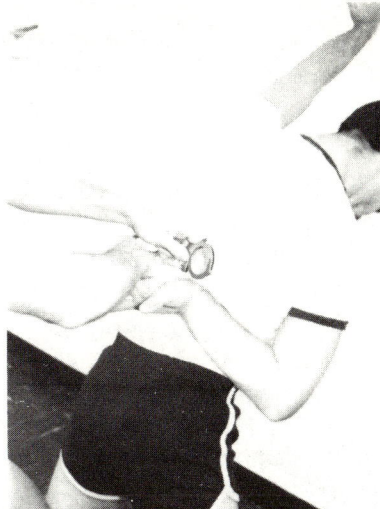

Figure 12-11

This pressure, assisted by gravity, should allow the single strand to actually swing free, enclose the wrist, and reattach itself within the ratchet mechanism. (See Fig. 12-12)

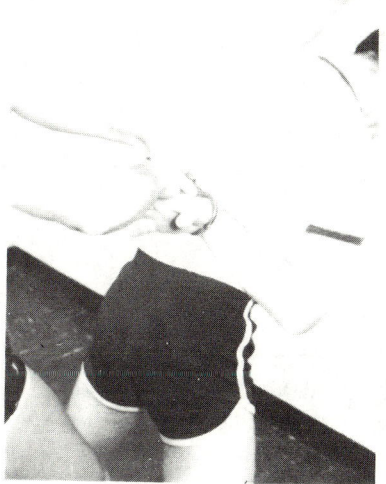

Figure 12-12

The right hand now grasps the total handcuff and aligns it properly between the back of the hand and directly in front of the prominent wrist bone. (See Fig. 12-13)

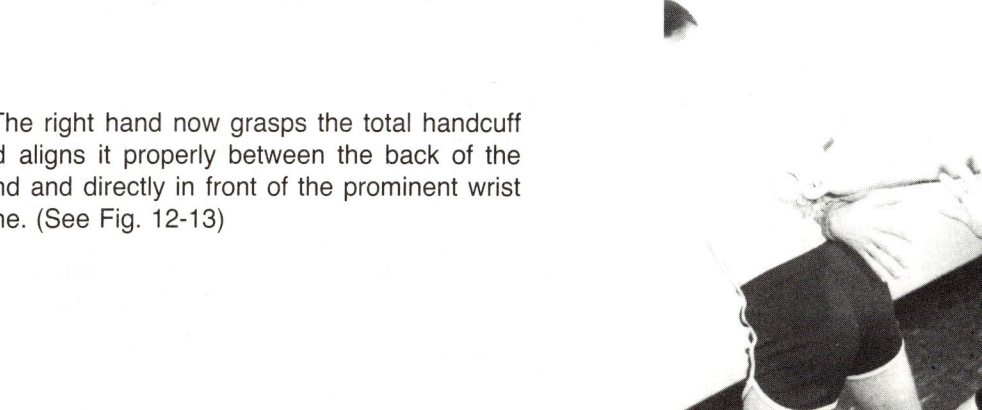

Figure 12-13

The handcuffs should be tightened snugly and securely, ever vigilant not to collapse the handcuff on the wrist so as to cause tissue or nerve damage.

Once the handcuff is secured, move your right hand down and grasp the handcuffs in the following manner. The right hand palm should lay across the chain of the handcuffs with the edge of the hand in contact control of the right handcuff and the thumb and index finger holding the left handcuff. (See Fig. 12-14)

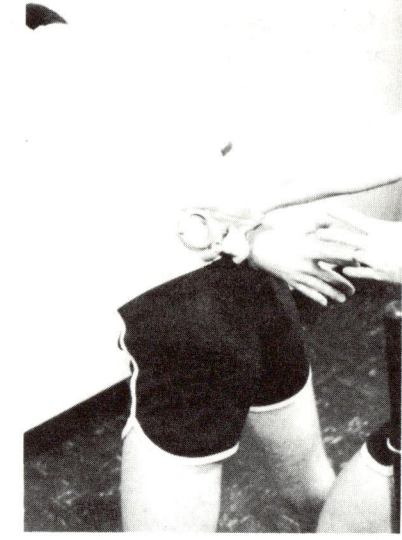

Figure 12-14

The backside of the officer's right hand should be positioned against the backside of the subject's right, handcuffed hand. (See Fig. 12-15)

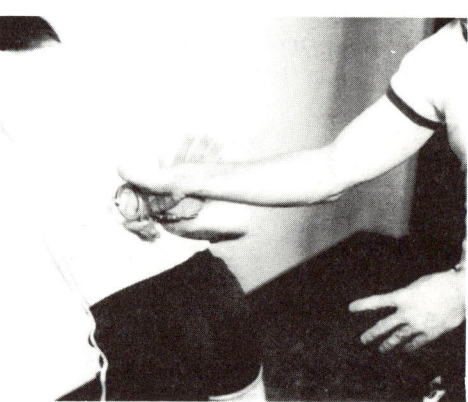

Figure 12-15

This allows for the officer to gain a more effective restraint advantage against the controlled wrist via both the handcuff band and the officer's induced back pressure on the subject's wrist. The right hand edge should lay against the major handcuff housing now on the subject's wrist, the handcuff chain should flow across the right palm of the officer, and the thumb and index fingers of the officer's right hand holds the handcuff in preparation of its application on the left hand. This left handcuff should have the double strand up, single strand down, with the lock opening port outward toward the officer.

Next, advise the subject to place his forehead on the wall and simultaneously bring his left hand back to the small of his back. (See Fig. 12-16)

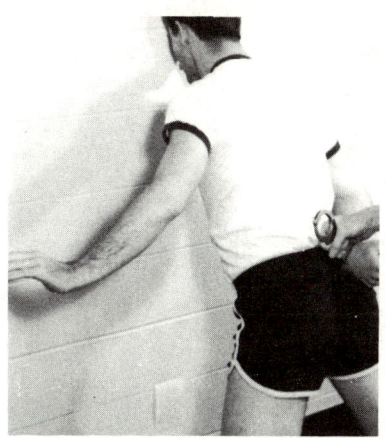

Figure 12-16

If support structure presents a safety risk as the subject appears somewhat incapacitated (drugs, alcohol, etc.), he may be moved into a Chest Compression Option position, which will be detailed later on in the Chapter.

The officer's free left hand should grasp the subject's left hand index and middle fingers, extend them appropriately, position the left wrist for eventual application of left handcuff. (See Fig. 12-17)

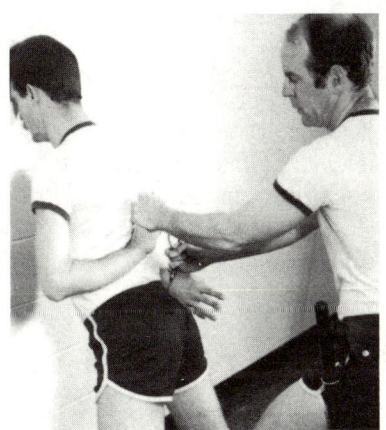

Figure 12-17

Again, move the handcuff to a point above the subject's left wrist applying pressure against the single strand. (See Fig. 12-18)

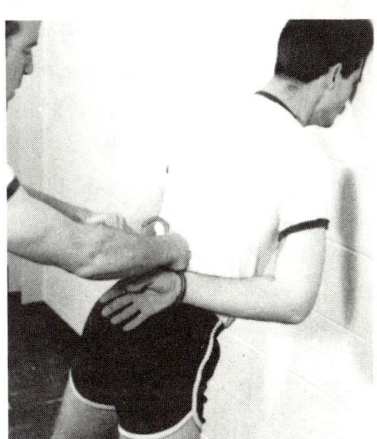

Figure 12-18

The single strand swings free and reattaches itself securing the left wrist. (See Fig. 12-19)

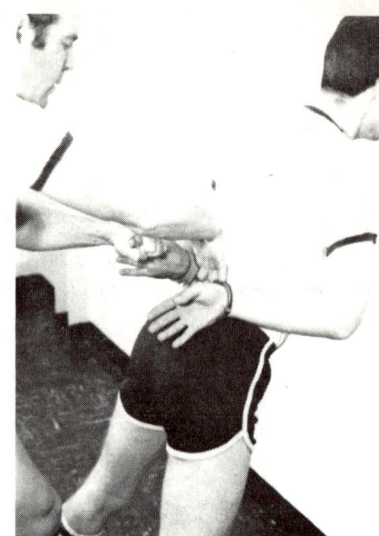

Figure 12-19

A visual check should reestablish the proper handcuff position sought; subject's hands together with palms out, handcuffs so situated with double strand up, single strand down, and lock mechanism ports out toward the officer. (See Fig. 12-20)

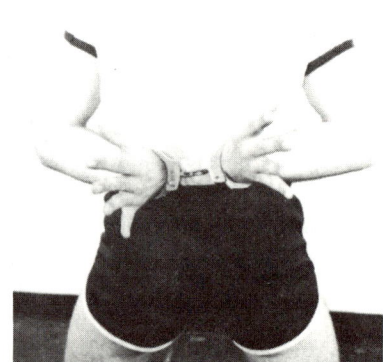

Figure 12-20

The officer now pushes the double-lock tab inward preventing further travel of the single strand. (See Fig. 12-21)

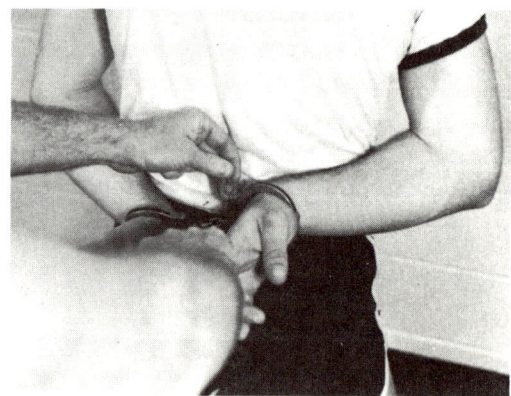

Figure 12-21

Handcuff Removal Procedures

The safest and most effective techniques for handcuff removal simply employs a reverse of their original application.

First, advise the subject that he is to heed your commands directly and with an immediate response. Advise him that if he does not understand the command, he may ask for clarification but must not *move.* In this exchange mode, unnecessary overreaction is controlled as both participants have mirrored expectations of action.

Next, have the subject place his forehead on the wall and ease his legs and feet as far back as necessary. (See Fig. 12-22)

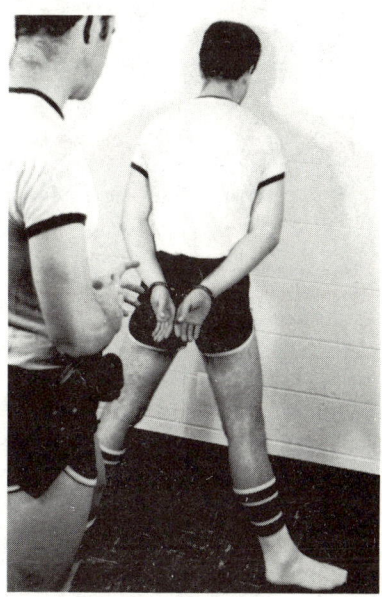

Figure 12-22

His feet should remain parallel to the wall and his overall body position should reveal obvious instability. As before, you remain at a safe distance directly to the rear of the subject in a proper search stance.

Now, grasp the index and middle fingers on the subject's right hand with your right hand. Twist the fingers upward and inward to provide the necessary state of tension. Once control has been established, remove the handcuff key from its secured site on your person. The handcuff key should be solitary on a key chain so that other keys do not inhibit its efficiency in entering the lock opening port, unlocking the double-lock device and releasing the handcuff. (See Fig. 12-23 and 12-24)

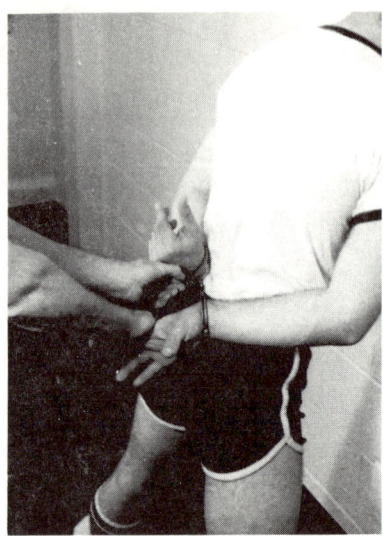

Figure 12-23

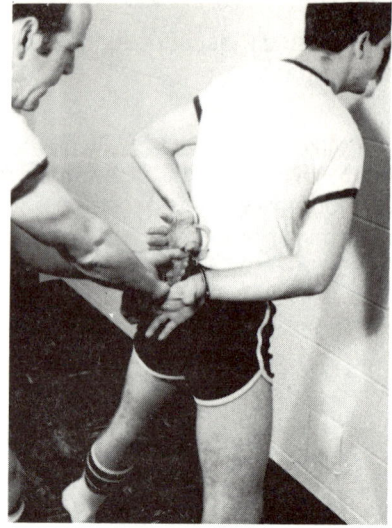

Figure 12-24

The subject is advised, previous to the release, to place his left hand, once released, onto the wall and now with this new means of support to slide his feet into a proper search stance. (See Fig. 12-25)

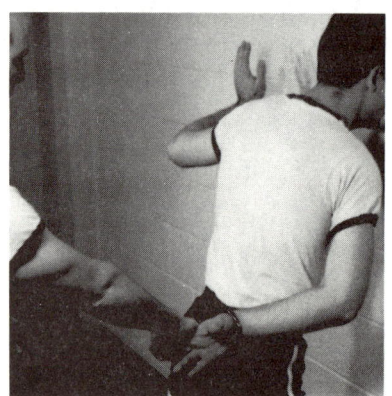

Figure 12-25

Reposition the loose single strand of the handcuff into the ratchet mechanism and now grasp the free handcuff chain and housing of right handcuff with your left hand. The subject's remaining right wrist is now controlled via the free handcuff and chain with the officer's left hand. (See Fig. 12-26)

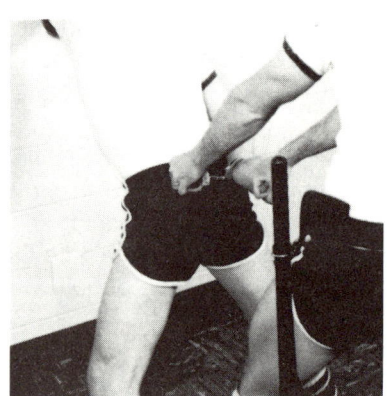

Figure 12-26

Advise the subject that upon release of his right wrist to place his hand onto the wall. With the key in your right hand, insert the key into the lock opening port, back-off the double lock status, and release the handcuff from the subject's right wrist. (See Fig. 12-27 through 12-29)

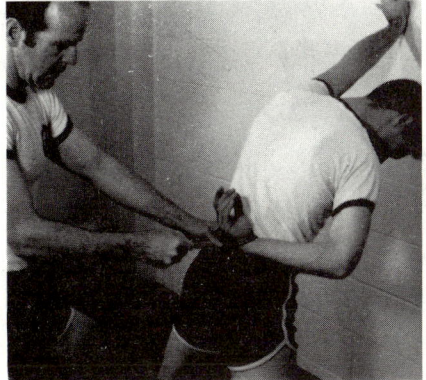
Figure 12-27

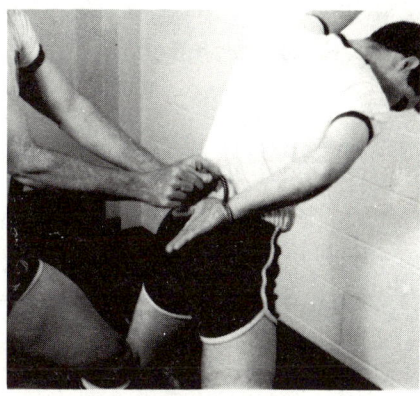
Figure 12-28

Figure 12-29

Once the right handcuff is free, reposition the loose single strand into the ratchet housing. Replace the handcuffs immediately back into your handcuff case and free both hands for potential control capability. (See Fig. 12-30 and 12-31)

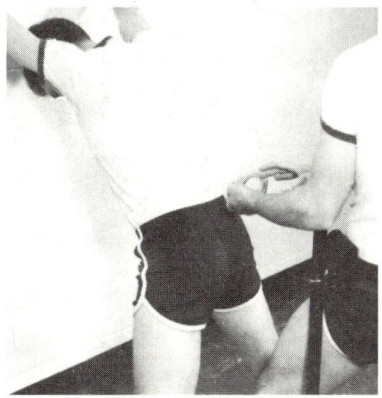
Figure 12-30

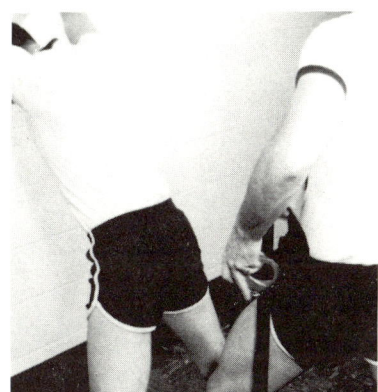
Figure 12-31

It is suggested that the subject be searched again once inside the security area to protect the officer from items the subject may have seized during transporting, and also to reinstill a no-nonsense, order response mental set in the mind of the suspect by the officer in this new environment.

Chest Compression Option

As noted previously, an alternative basic procedure is provided once the subject's right hand is handcuffed. The officer may choose to command and gradually push the subject's chest first against the wall via the officer's left elbow pushing against the subject's back, while pulling upward on his belt, etc. (See Fig. 12-32)

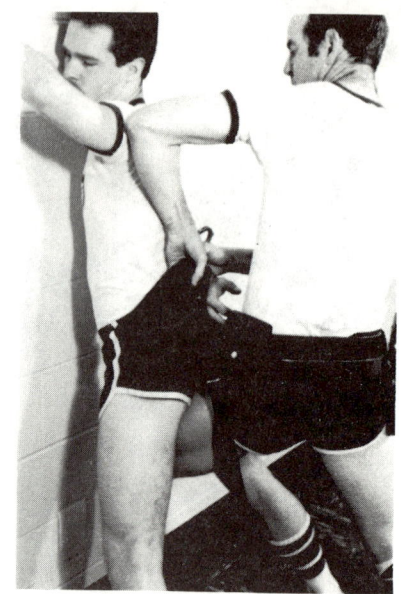

Figure 12-32

The officer's forward leg can apply up-lift pressure against the subject's hip and groin region. (See Fig. 12-33)

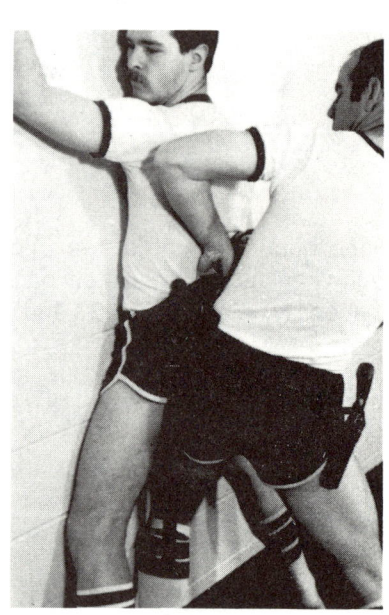

Figure 12-33

Once this pinning pressure is maximized, the subject is commanded to keep his head and face away from the officer. He now is advised to slowly move his left hand to the small of his back for eventual handcuffing. (See Fig. 12-34)

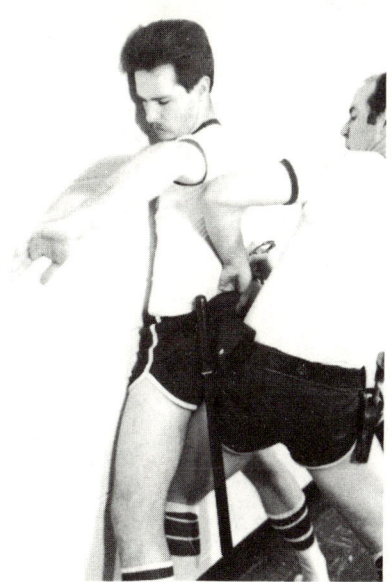

Figure 12-34

The officer again grasps the weak hand appropriately and systematically applies the left handcuff on the subject. (See Fig. 12-35 and 12-36)

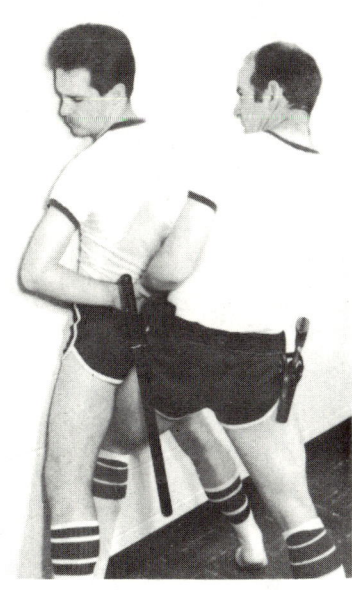

Figure 12-35

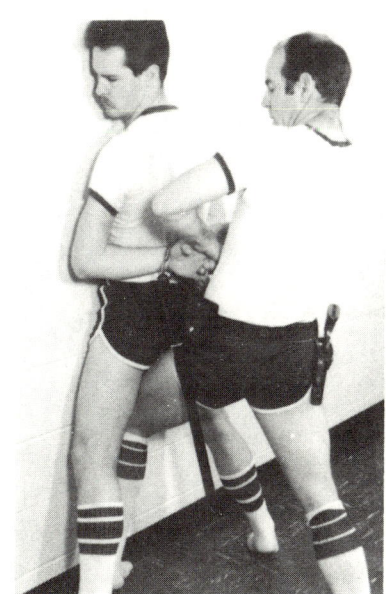

Figure 12-36

Chapter 13

SPECIALIZED SEARCH TECHNIQUES

Prone Search/Standing Option

Since the prone technique mentioned limits the accessibility of the officer into body areas where potential weapons, contraband, etc. can be secreted, a modification must be incorporated. This modification initially begins with the subject being handcuffed in the prone position via either the Leg Tuck or Arm Pin technique mentioned in the following chapter. Whichever of the two techniques are incorporated, it is presumed that handcuffs have been properly applied.

Subsequently, the subject is assisted to his feet by either wrist control or support rendered to the subject's upper arm area via a grasp or actual arm lock maneuver. (See Fig. 13-1 and 13-2)

Figure 13-1

Figure 13-2

Once on his feet, the subject is instructed to spread his feet approximately shoulder length apart, pivot on his heels, and come to a stance where both feet have toes pointed outward. (See Fig. 13-3)

Figure 13-3

141

In this position, the arrested subject has minimal balance and the officer is made less vulnerable to a rear or side kick attack. The officer now approaches the presumed strong side of the subject (since the handcuffing sequence has already concluded), and it is the side most likely to evidence a hurriedly, secreted weapon. On the initial approach, the officer's left leg is moved between the leg arch of the suspect with the officer's left foot extended through theh arch so that the inner thigh of the officer's left leg is in contact with the interior upper leg/buttocks region of the subject. (See Fig. 13-4)

Figure 13-4

Concurrent with this activity, the officer either grips or preferably places his nightstick down over the handcuff chain between the handcuffs and the subject's buttocks area. (See Fig. 13-5)

Figure 13-5

The tip of the stick is placed into the tailbone area and pressure is now applied at the grip end of the stick so that the subject's upper torso is bent backward increasing the unbalanced state of the subject and drawing him closer to the officer for searching activity. (See Fig. 13-6)

Figure 13-6

After the subject is extended backward in association with the baton leverage, the officer can now do a full upper and lower torso search of the subject. (See Fig. 13-7)

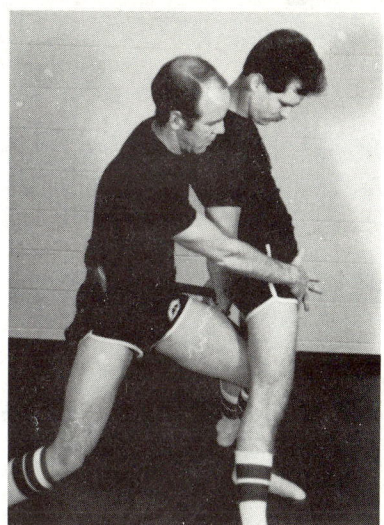

Figure 13-7

Once the right side has been searched the officer exchanges hand control on the grip of the baton maintaining the backward extension of the suspect. The officer moves his right leg into the leg arch of the subject with the right foot planted beyond the arch, with the officer's upper, interior right thigh against the subject's interior left thigh. The search now commences with the officer's left hand searching the upper and lower left torso of the subject. (See Fig. 13-8)

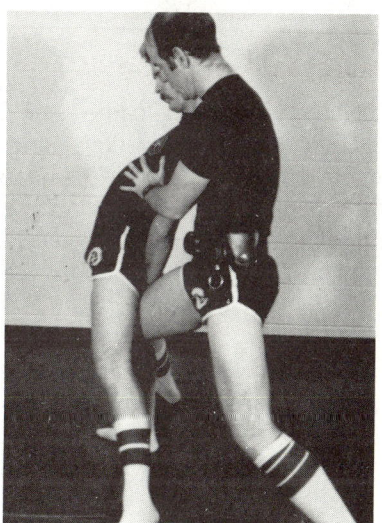

Figure 13-8

Kneeling Search

Cautiously, talk the subject into a kneeling position with his hands (fingers) interlaced on top of his head, crossing his left ankle over his right. During this initial contact, the officer may choose to assume an enhanced defensive position via a baton "on guard" stance. (See Fig. 13-9)

Figure 13-9

Either during the directorial conversation or after he has reached the concluded kneeling position, the officer should gain a position to the rear of the subject. (See Fig. 13-10)

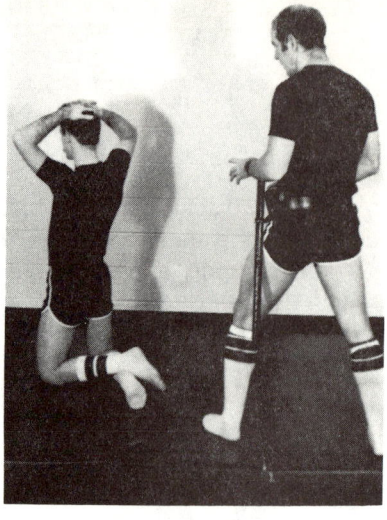

Figure 13-10

Initially, move up to the rear of the suspect reaching up with your right hand, slipping your fingers between the subject's head and his interlaced fingers. Establish a firm grasp of the subject's fingers and hair with sufficient pressure to negate either vertical or horizontal escape. (See Fig. 13-11)

Figure 13-11

With this grip established, the officer's right knee is placed in contact with the rear, lower back of the subject. With the grip base and the pressure of the knee in the lower back of the suspect, extend him backward to increase his unbalanced state. (See Fig. 13-12)

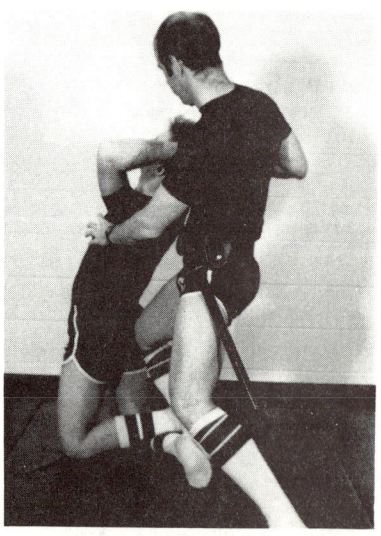

Figure 13-12

The officer now plants his left foot in a strong, extended stance and commences a full upper and lower torso search with his left hand. (See Fig. 13-13)

Figure 13-13

Once the left upper and lower body is searched, the officer exchanges his firm grip on the subject's interlaced fingers by removing his right hand immediately resuming the control with his left hand. Simultaneously, the grasp of the subject's fingers is established, the officer moves his left foot and ankle between the crossed ankles of the subject and places his left knee into the rear lower back of the subject. (See Fig. 13-14)

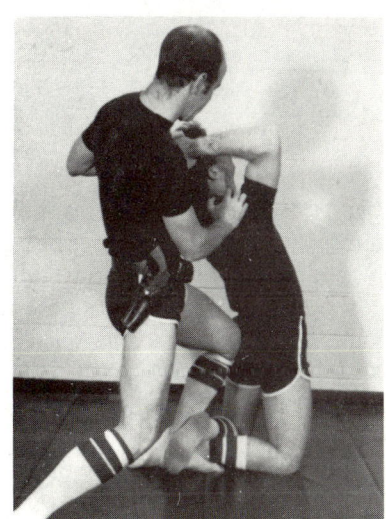

Figure 13-14

The same backward extended position is instituted and the subject's upper and lower body is thoroughly searched by the officer's right hand once he has established a stable, extended stance with the movement of his right foot.

Prone Sideroll Search

If the subject is prohibited from standing due to impairment or injury, he can be searched while still on the ground.

First, move to the presumed strong side of the subject. Grasp the subject's weak side shoulder and elbow area and roll him on to his strong side. (See Fig. 13-15)

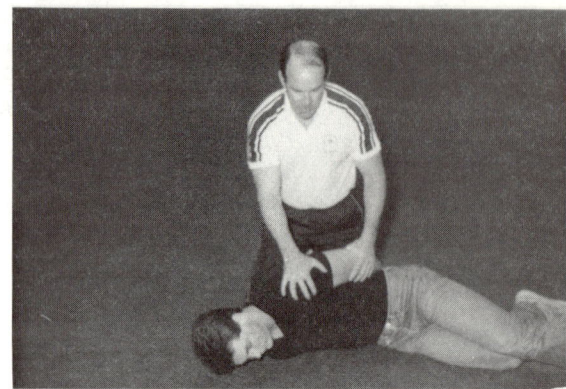

Figure 13-15

The officer can assume a kneeling position to the rear of the subject's upper back, commencing to search with the weak hand, stabilizing the subject's position with his strong hand. The officer should make sure that his entire body remains above the hands to assure safety. (See Fig. 13-16)

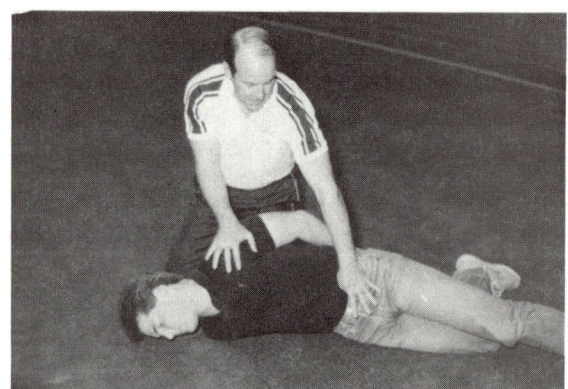

Figure 13-16

Once the search of the weak side is complete the officer now pivots around the head of the subject away from the subject's feet and legs. (See Fig. 13-17)

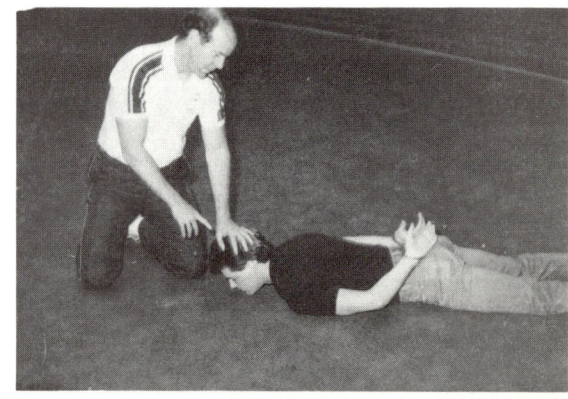

Figure 13-17

The officer now assumes the same kneeling position to the rear of the subject's upper back. He now grasps the subject's strong side shoulder and elbow area and rolls him onto his weak side. (See Fig. 13-18)

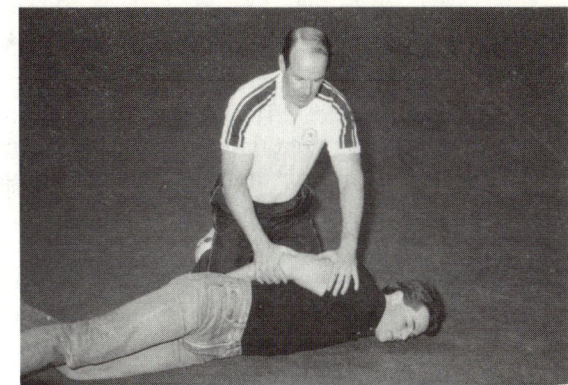

Figure 13-18

The officer now commences the search of the subject's strong side, stabilizing the subject's position with his weak hand while searching with the strong hand. Again, safe positioning must always remain above the subject's hands. (See Fig. 13-19)

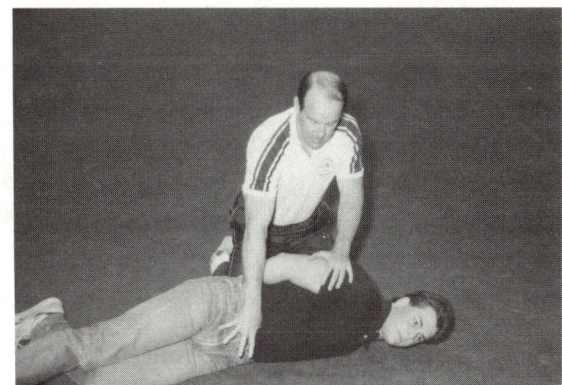

Figure 13-19

Once the search has been completed the subject is assisted to his feet by techniques previously demonstrated.

Frisk Technique

Courts have held that an officer can initiate a "pat down" or "frisk" of a subject if sufficient threat potential does exist to the officer during a street stop.

Obviously, if the subject will move cooperatively into the standard wall search position and a supportive structure is available it presents the most advantageous position for the officer.

However, if both factors do not exist, an alternative technique does allow the officer to perfect a cursory search while keeping the subject in a disadvantaged position.

Move to the presumed strong side of the subject (position 10 o'clock). The officer reaches across his body with his right hand, grasping the fingers of the subject's right hand initiating a reverse gooseneck technique thus hyper-extending the right arm.

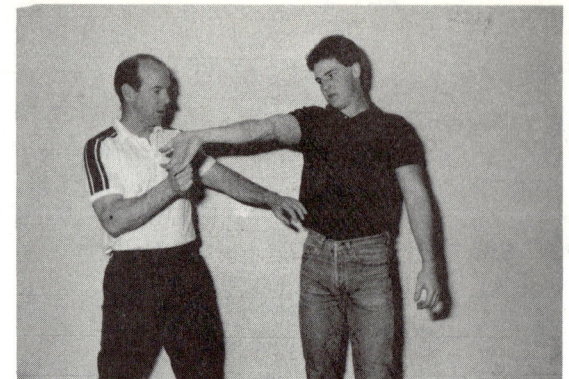

Figure 13-20

The officer now conducts a strong side, upper torso frisk of the subject with his left hand.

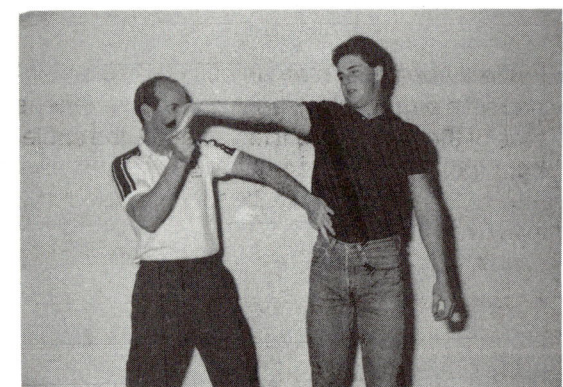

Figure 13-21

Upon completion the officer moves behind the subject maintaining vigilance and distance.

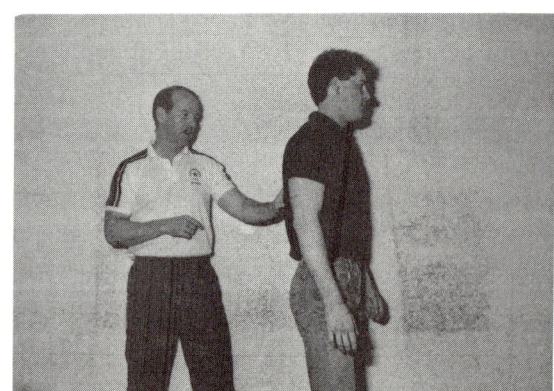

Figure 13-22

The officer moves to the presumed weak side (2 o/clock) position of the subject. He now initiates the grasp of the subject's left hand with his left hand and establishes the reverse gooseneck technique on the hand hyper-extending the affected arm. He then conducts the weak side, upper torso frisk with his right hand.

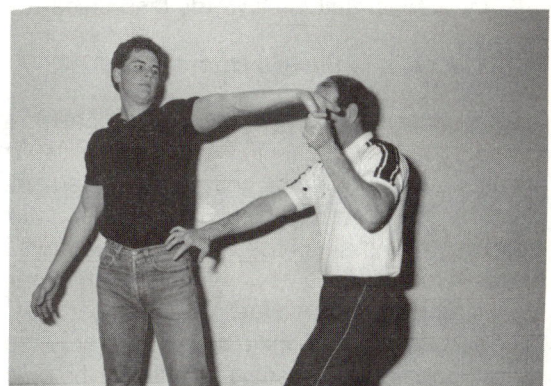

Figure 13-23

Female Searches

The search of a female presents special problems to field personnel. Obviously, when conditions allow same sex searches it appears to be the most effective and recommended search regimen.

However, an arrest situation may necessitate male searches of females and therefore, a sound procedure must be established. Remember the same criteria for a search following an arrest hold true for men or women, detection of hidden weapons or contraband and the securing of a situation of safety for both parties. The female suspect can in fact be considered more dangerous than the male, not only since the immediate physical threat can never be under-estimated, but the potential for subsequent civil, criminal, or administrative charges of abuse are of increased concern.

It should be noted that many criminal groups utilize the women members to carry weapons and other illegal paraphernalia, i.e.: drug dealers, outlaw motorcycle gangs, street gangs, etc. This is done because of a common belief that during an arrest situation, the male police officer will not search a female as thoroughly or possibly neglect searching the female at all. The male suspect will appear or in reality be "clean" while the female counterpart possesses the weapon and/or contraband of the criminal activity. In this situation the officer must exercise added caution because the female is expected to use the weapon against the police officer or pass the weapon to the male suspect if in fact she is being prepared for search and arrest.

If department policy permits the procedure and there is no access to a female police officer, the male officer has to be concerned with five (5) major areas of female searches:

   a. The person
   b. The outer garments and clothing
   c. The purse
   d. The shoes or boots
   e. The hair/wig

The basic guide lines for searching females are simple:

1) A skillful and thorough check of the person while maintaining an absolute professional and impersonal attitude towards the subject.

2) A witness to the search of the suspect.

As a male officer, you have an obligation to search the female after the arrest and prior to transporting the prisoner in the patrol vehicle. The attitude of either gender of police officer in these circumstances should be completely impersonal, free of displays of nervousness, statements that might be offensive or subject to misinterpretation by the suspect. With the exception of emergency situations, the male police officer should have a witness to the search of a female prisoner. The witness would ideally be an independent, impartial female, however, if one is not readily available, a ranking officer or another police officer may serve as a witness to the physical search of the female suspect. If a female citizen is nearby, and consents to the request, she should oversee the entire period of officer/subject contact. The officer should clearly indicate the nature of his request to subject and anticipate those actions that might be misinterpreted by either party and preface his actions with design dialogue. The officer additionally should record sufficient identification of the women assisting in the task of overseeing the searching activity.

Since it is not beyond the scope of the female criminal to conceal a weapon, such as razor blades, drugs, etc., under the folds of the breasts or vaginal or rectal orifices of the body, a more detailed clinical search of the female prisoner is necessary so long as the offense warrants such action. Once the street search of the female has concluded, the suspect should be transported to the station or other functional facility (hospital, clinic, etc.) where a qualified female can conduct the strip search.

It should be noted that emphasis should be placed on the fact of femininity secondarily while qualifications and tactical searching skills must be primary. This person must be trained in order to discover evidentiary material, perceive potential weapons present, and be capable of processing skills (report writing, evidentiary logging, and subsequent court testimony) if necessary.

If the female is wearing a coat and weather permits, have the suspect open the front to expose the lining, then have her re-face the wall. The next procedure is for the officer to hold the coat by the rear collar and have the suspect step forward and out of the coat. The whole coat may be examined after the physical search of the suspect. The coat may be returned to the suspect prior to handcuffing if the weather is too cold for being without an outer garment.

The remaining outer garments worn by the female are searched similarly to those worn by the male suspects: the fabric is squeezed very carefully and methodically to determine the presence of weapons or contraband. If an article is discovered on the person or in the clothing of the female, removal is imperative for the safety of the officer.

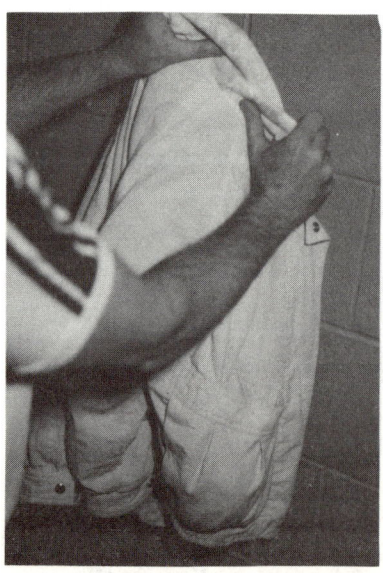

Figure 13-24

Removing weapons or illegal contraband from the female suspect must be done in a completely impersonal and unobtrusive manner. The reflected professional attitude of the searching police officer will determine the cooperativeness of the suspect and reduce the repercussion of searching a suspect of the opposite sex.

If a weapon of obvious danger to the officer is located in the bra of the suspect, the officer may unsnap the bra from the rear and allow the weapon to fall from concealment. Or if the weapon is small in nature, such as a razor blade or small knife, and is exposed at the top of the bra, the officer may reach into the neck opening of the blouse and extract the weapon using the thumb and index finger of his hand. The officer should refrain from entering the neck opening with the full hand or reaching lower than the top of the bra.

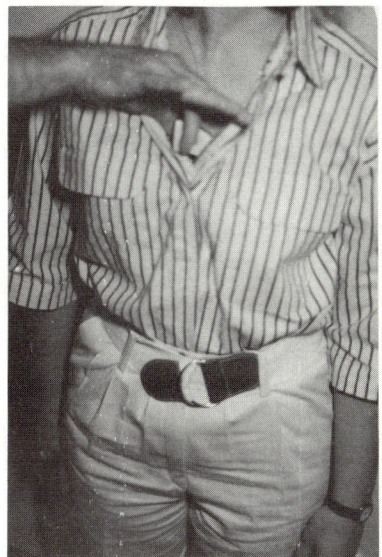

Figure 13-25

Any weapons located by the officer in the waist or belt line of the suspect must be removed and secured using the same professionally discreet manner outlined previously.

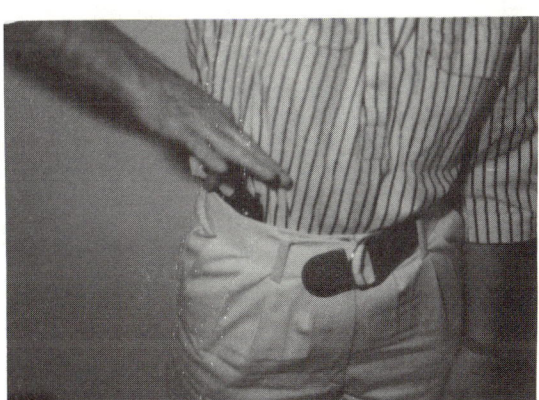

Figure 13-26

Long or thick hair poses a problem of concealment for searching by the officer. The hair must be carefully searched while being cautious for the presence of razor blades, needles or other small, sharp objects typically carried in the hair or wigs of female subjects. All wigs should be removed and examined for the hiding of weapons or contraband under the false scalp. The wig may be retained as evidence without being returned to the suspect at the scene.

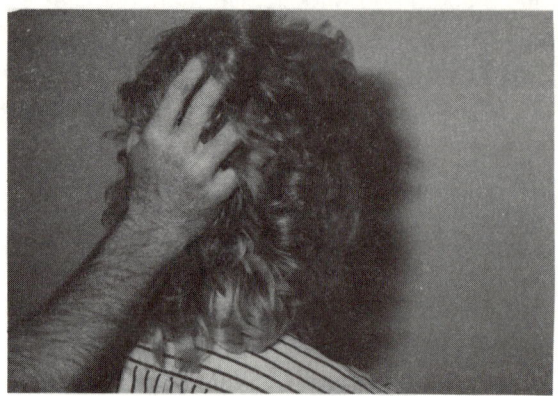

Figure 13-27

Remember, although the pattern of searching should be as systematic as that of the male; the breast, crotch and buttocks areas have only experienced a cursory search in the field. Therefore, during escort and transport, and other processing steps, each officer must maintain the highest degree of vigilance toward the female subject.

Chapter 14

SPECIALIZED HANDCUFFING TECHNIQUES

Prone Position

Leg Tuck

In a safe, controlled manner, talk the subject into a prone position. Gain a safe position to the rear of the subject and commence to instruct the subject in an authoritative manner, the following controlling commands. (See Fig. 14-1)

Figure 14-1

First, the subject should be advised to raise his left leg and place the left ankle over the back of his right knee, and additionally rotate his palms upward. (See Fig. 14-2)

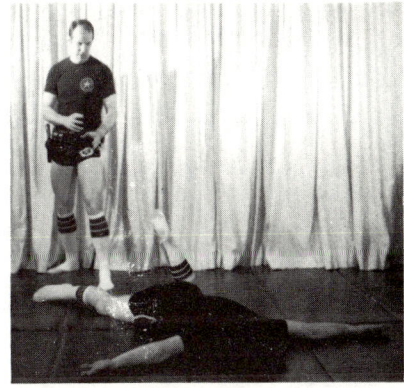

Figure 14-2

The subject should then be told to lift his right lower leg and contact the leg, trapping his left ankle in the process. (See Figs. 14-3, 14-4)

Figure 14-3

Figure 14-4

153

Simultaneously, the officer now cautiously approaches the subject, straddling the tucked right leg adding additional controlling weight and compression to the now immobilized lower torso. (See Fig. 14-5)

Figure 14-5

It should be noted that the officer should hold his hands in a groin block posture until the inner surface of the officer's right leg contacts the subject's right foot. (See Fig. 14-6)

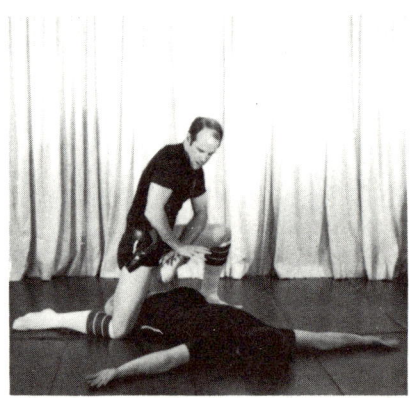

Figure 14-6

In this position, the subject now can be safely handcuffed in the standardized style studied earlier. (See Figs. 14-7, 14-8)

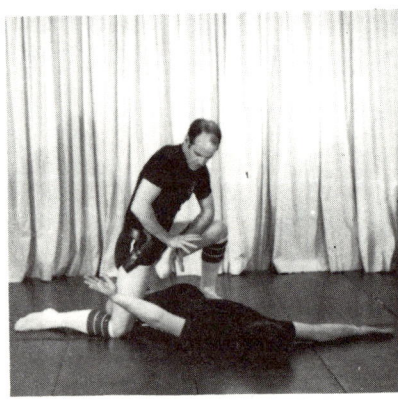

Figure 14-7

Figure 14-8

If during the handcuffing process, the subject were to become unruly, the officer can remain in this straddled position and reach forward and grasp the clothing or if necessary the hair of the subject to increase the mode of stabilization as now both the upper and lower torso experience increased tension.

Arm Pin

In a safe, controlled manner, talk the subject into a prone position. Once prone, the officer should assume a position off to the left (presumed weak side) of the subject. (See Fig. 14-9)

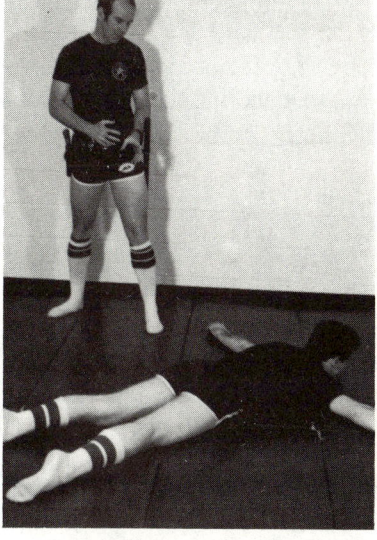

Figure 14-9

With concise, authoritarian commands, instruct the subject into a more controlled pose. First, advise the subject to turn his head to the right. Next, his right hand should be fully extended to the side with palm up. He should then be advised to bend his left arm and place his left hand, palm up, on the small of his back. (See Fig. 14-10)

Figure 14-10

Now, the officer should continue to approach from the left side. The officer's right knee now should be positioned in contact with the back of the subject's left elbow while the officer's hands apply pressure to the back of the subject's left hand, increasing tension on the wrist. (See Fig. 14-11) The left arm is now stabilized at the elbow by the officer's right knee and lower thigh and by pressure application to the subject's left wrist.

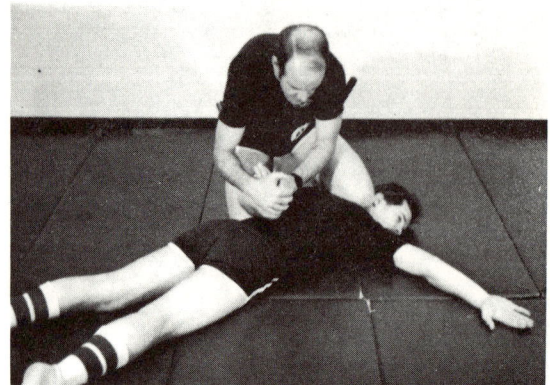

Figure 14-11

Once the presumed weak hand is controlled, the officer should place his left knee along the side of the subject's right ear and mastoid region of the neck. (See Fig. 14-12)

Figure 14-12

The weight of the officer via the knee contact, and the pressure to the mastoid area of the neck will debilitate and control the subject. Obvious caution must be present so as not to cause a collapse of the subject's carotid artery.

Now controlled, the subject is advised to bring his right hand to the small of his back with the palm up. (See Fig. 14-13)

Figure 14-13

The officer now reaches with his left hand for his handcuffs, places the handcuffs face up on the subject's back, and begins to apply the handcuff on the subject's right wrist while the officer's right hand still exerts pressure on the subject's left wrist. Finally, the left handcuff is applied to the subject's left wrist. (See Fig. 14-14, Fig. 14-15, Fig. 14-16, and Fig. 14-17)

Figure 14-14

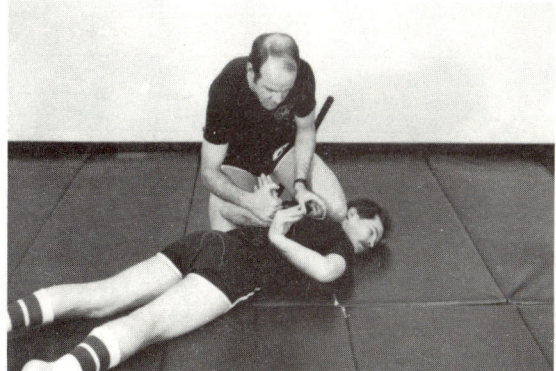

Figure 14-15

Figure 14-16

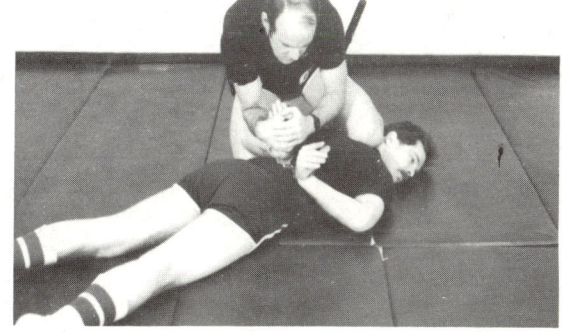

Figure 14-17

Kneeling Position

With a few minor variations, the kneeling handcuff technique is almost parallel with the standard handcuffing style extensively discussed in an earlier lesson. After stepping clear of the subject following the complete right upper and lower body search, the officer should now be in his alert stance directly behind the kneeling suspect. (See Fig. 14-18)

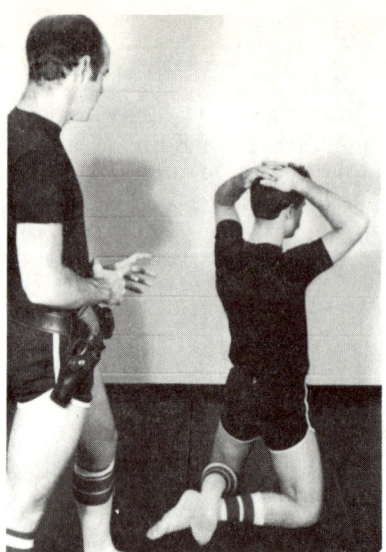

Figure 14-18

The officer advises the subject to keep his hands laced together on top of his head and re-approaches the subject placing his left foot and ankle between the still crossed ankles of the subject. (See Fig. 14-19)

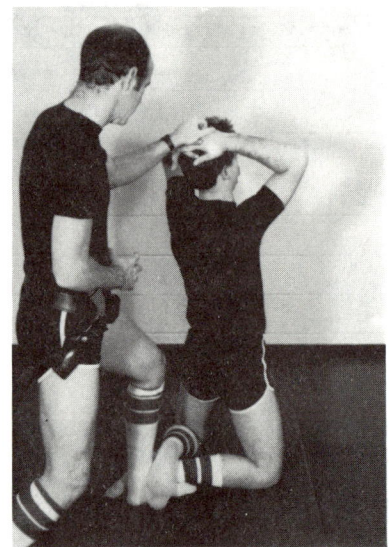

Figure 14-19

Once establishing control of the subject's laced hands with his left hand, the officer advises the subject to bring his right hand back to the small of his back while keeping the left hand controlled on the subject's head. (See Fig. 14-20)

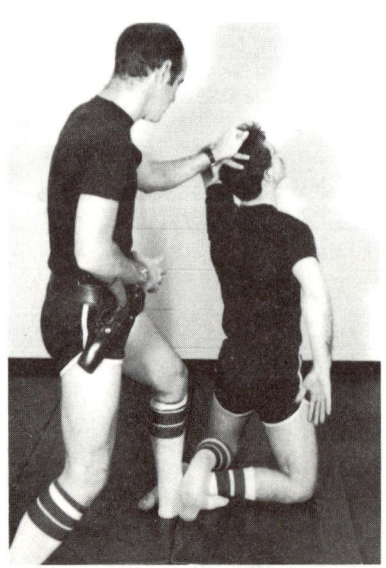

Figure 14-20

158

The officer establishes a firm grip on the right hand, removes his handcuffs from his holder, and applies the handcuff to the right wrist. (See Fig. 14-21 and Fig. 14-22) Too much risk of injury as is. Need to push subject forward prior to any handcuffing or left elbow is free to attack.

Figure 14-21

Figure 14-22

The officer has the option of allowing the subject to remain in a verticle kneeling position for the handcuffing mode or pushing the subject forward for even greater debilitating effect. (See Fig. 14-23)

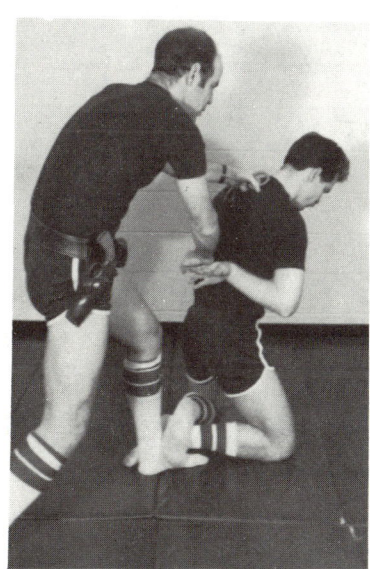

Figure 14-23

The officer proceeds to move his hand in a position to exert pressure against the subject's wrist and applied handcuff in readiness for the application of the left handcuff. (See Fig. 14-24)

Figure 14-24

The officer then orders the left hand to be brought to the rear and placed in the small of the back and eventually being grasped by the officer's left hand. The left handcuff is applied and the handcuffs are double locked. (See Fig. 14-25, Fig. 14-26 and Fig. 14-27)

Figure 14-25

Figure 14-26

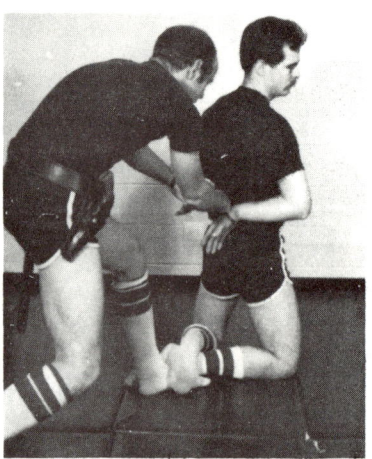

Figure 14-27

Wall Compression Technique

Advise the subject to face the wall with both hands spread and placed high on the wall. (See Fig. 14-28)

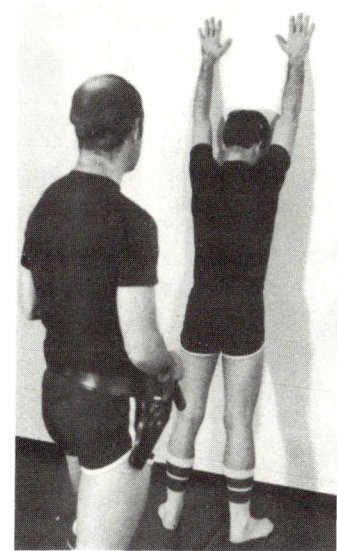

Figure 14-28

The subject is then told to slowly move toward the wall and upon contact, proceed to turn his toes outward so that his heels and interior ankles are touching the wall. (See Fig. 14-29)

Figure 14-29

The subject will experience difficulty in maintaining the wall position; therefore, a gravity center balance can be provided by the officer with his fist or baton tip.

The subject is told to place his left hand to the rear of his head while looking to the left. (See Fig. 14-30)

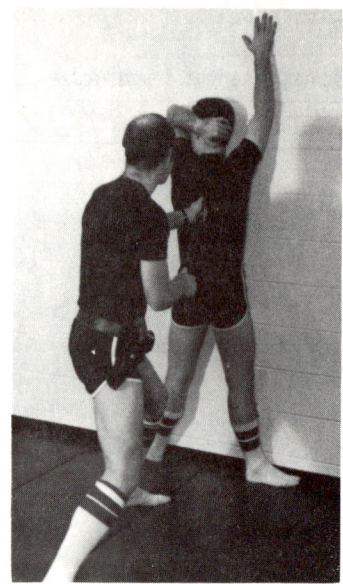

Figure 14-30

Next, the officer commands the subject to place his right hand to the small of his back with the fingers pointed outward toward the officer. (See Fig. 14-31)

Figure 14-31

The officer should perceive that the subject will have difficulty maintaining any degree of balance since the wall is actually working against balance acquisition. At this point, the officer grasps the subject's right hand with his left hand and blocks the right elbow with his right hand. (See Fig. 14-32 and Fig. 14-33)

Figure 14-32

Figure 14-33

Simultaneously, the officer compresses the subject to the wall with his right leg pushing up and in at the subject's buttocks and then quickly moves to a rear wrist lock on the suspect. (See Fig. 14-34 and Fig. 14-35)

Figure 14-34

Figure 14-35

The officer increases the tension on the wrist joint and tilts the right shoulder of the subject to the right side increasing the instability of the subject. (See Fig. 14-36)

Figure 14-36

The officer removes his handcuffs from the handcuff case and applies the handcuff with his left hand. (See Fig. 14-37)

Figure 14-37

To minimize the opportunity to injure the eye or face of the suspect, the officer should push upward with the handcuff as it is applied to the subject's left wrist rather than an over hand application where the single strand could actually strike the suspect. The subject's handcuffed left wrist is moved down to the rear lower back while the right wrist and arm remains in its trapped position. (See Fig. 14-38)

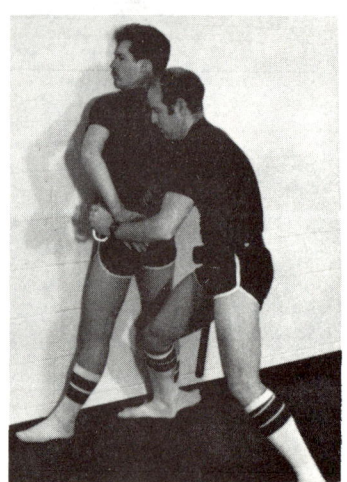

Figure 14-38

The free handcuff is looped under the secured left wrist providing a more accessable position for subsequent right wrist application. (See Fig. 14-39, Fig. 14-40 and Fig. 14-41)

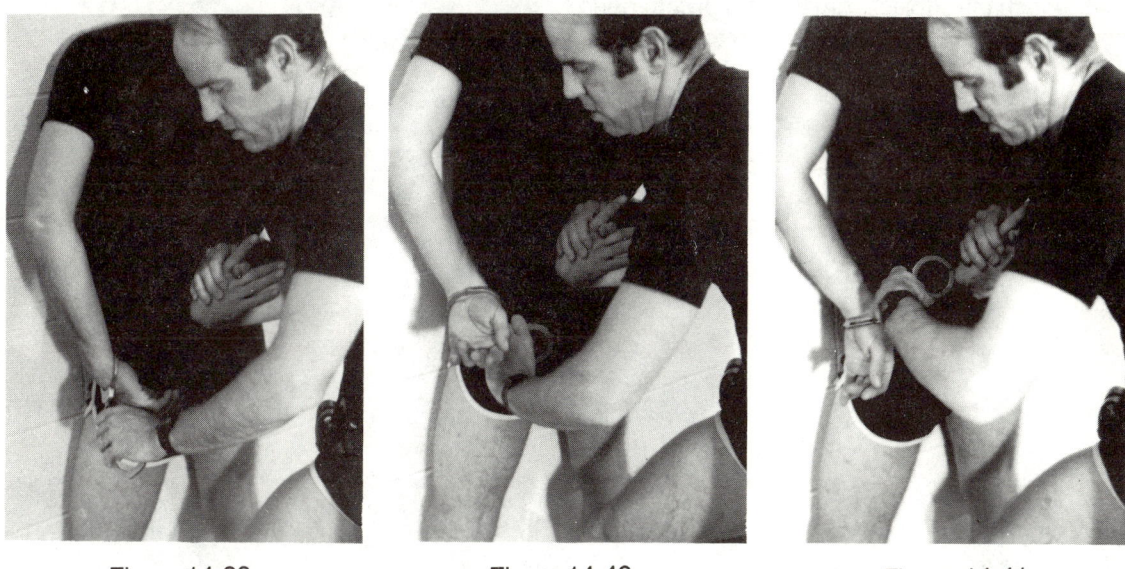

Figure 14-39 Figure 14-40 Figure 14-41

The right wrist is now handcuffed while the secured condition is maintained. (See Fig. 14-42 and Fig. 14-43)

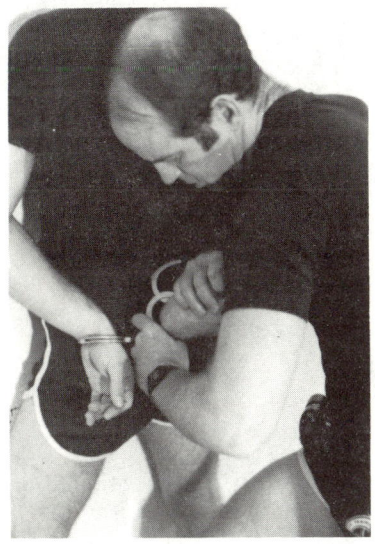 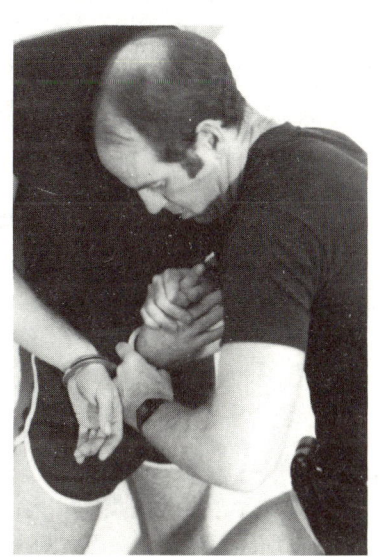

Figure 14-42 Figure 14-43

Standing Subject

Approach the subject from the rear of command the subject to turn away with his back to the rear.

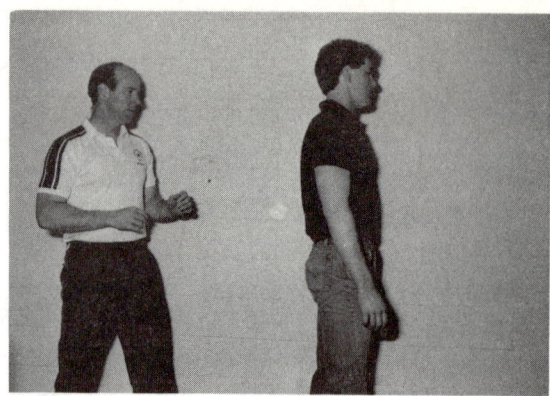

Figure 14-44

Advise the subject to raise his hands high in the air with palms to the rear.

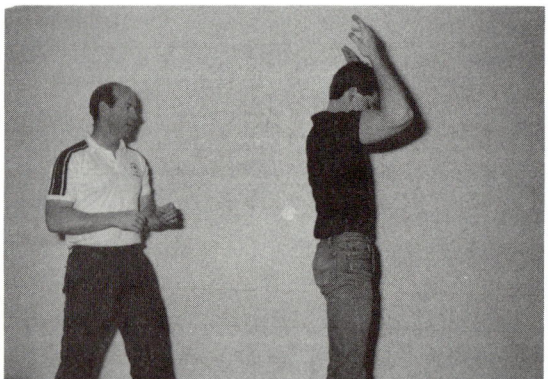

Figure 14-45

Have the subject spread his legs wide apart pointing his toes outward. Instruct the subject to bend forward while simultaneously moving his arms downward eventually bringing his hands to the rear in an effort to de-stabilize the individual.

Figure 14-46

Advise the subject to turn his plams outward and his thumbs up. His head should remain forward and upward. Move in and initiate the basic handcuffing technique on the strong, then weak hands.

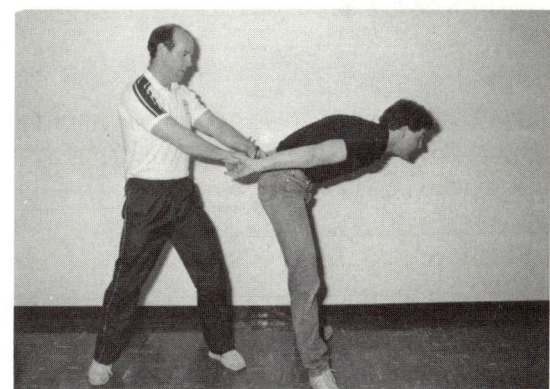

Figure 14-47

An optional technique can be initiated once the subject has assumed the hands overhead and toes pointed outward. Here the subject is ordered to bend backward, bringing his hands to the rear in this de-stabilized position for eventual handcuffing.

Figure 14-48

Cover Handcuffing

Circumstances may dictate that an officer handcuff a subject while obstructions exist. In such cases the officer may actually utilize these forms of situational cover as a mechanism of advantage and an effective strategy for safety. The basic principle in this procedure is to promote a tactical advantage for the officer by the use of cover (vehicles, doorways, other stable objects, etc.) to inhibit aspects of assaultive behavior on the part of the arrested subject.

We have listed a few typical applications of this approach to initiate your perception of the potential to the wide variety of street situations where its utilization may seem appropriate.

Vehicle Door (Driver)

The subject is ordered to place his left hand out of the window opening with the arm fully extended palm away from the vehicle. (See Fig. 14-49)

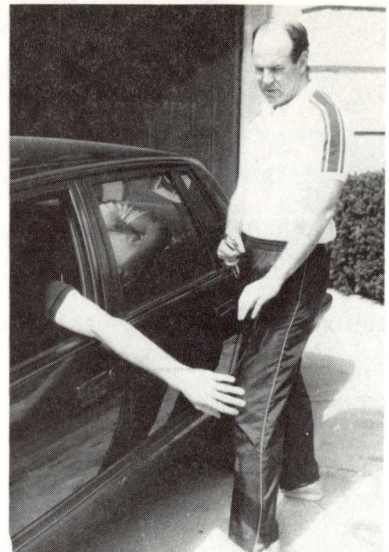

Figure 14-49

The officer secures the left hand and applies the handcuff. (See Fig. 14-50)

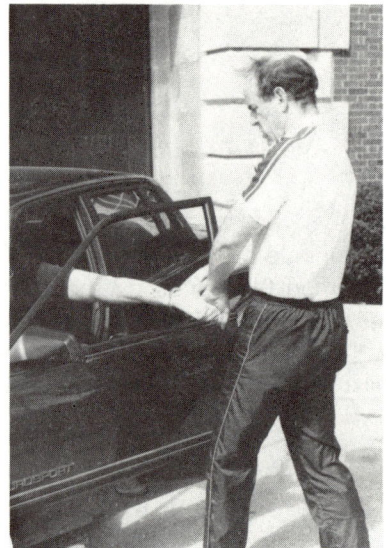

Figure 14-50

The officer maintains a secure grip on the hand via finger grasp and handcuff hold. The officer slowly opens the driver's door ordering the driver to exit from the vehicle. (See Fig. 14-51)

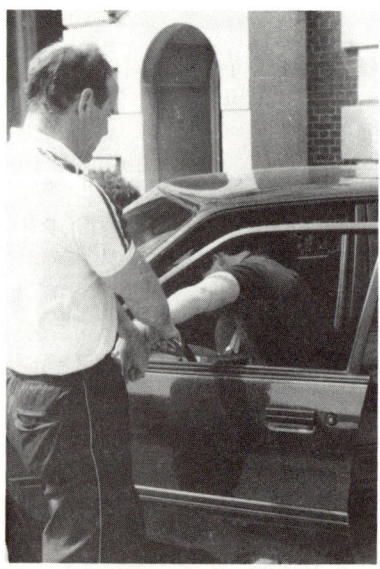

Figure 14-51

The subject is kept turned away from the officer and advised to place his right hand behind him and through the window where it is now handcuffed. (See Fig. 14-52)

Figure 14-52

The officer moves the subject away from the door and eventually into an escort position. (See Fig. 14-53)

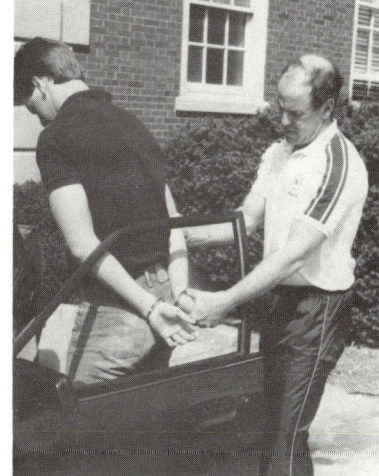

Figure 14-53

Vehicle Door (Passenger)

The subject is ordered to place his right hand out of the window opening with the arm fully extended, palm away from the vehicle. (See Fig. 14-54)

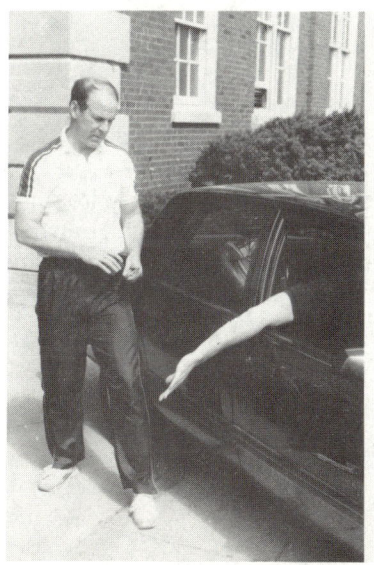

Figure 14-54

The officer secures the right hand and applies the handcuff. (See Fig. 14-55)

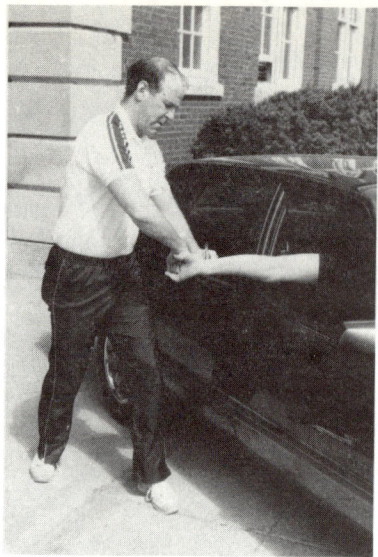

Figure 14-55

The officer maintains a secure grip on the hand via finger grasp and handcuff hold. The officer slowly opens the passenger door, ordering the passenger to exit from vehicle. (See Fig. 14-56)

Figure 14-56

The subject is kept turned away from the officer and advised to place his left hand behind him and through the window where it is now handcuffed. (See Fig. 14-57)

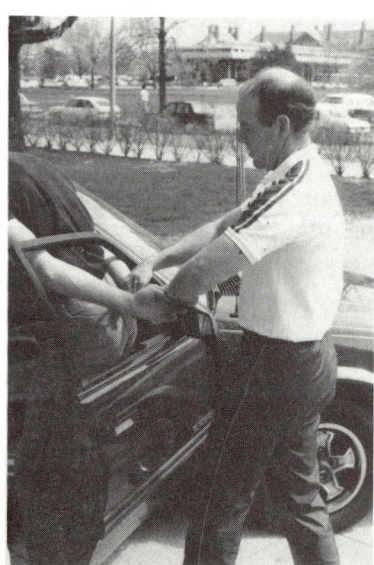

Figure 14-57

The officer now moves the subject away from the door and eventually into an escort position. (See Fig. 14-58)

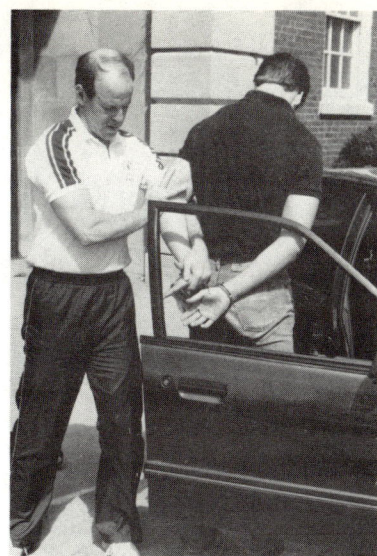

Figure 14-58

Vehicle Door (High Risk Traffic)

As a result of a high risk traffic stop, etc., and individual has been advised to walk backward toward the police vehicle. The subject is verbally positioned to a point directly in front of either the driver or passenger officer's door. (See Fig. 14-59)

Figure 14-59

The subject is now directed to slip his hands and arms through the window frame with the palms outward and thumbs up. (See Fig. 14-60)

Figure 14-60

The subject is now handcuffed with the officer remaining behind the cover of the door until escorting is begun.

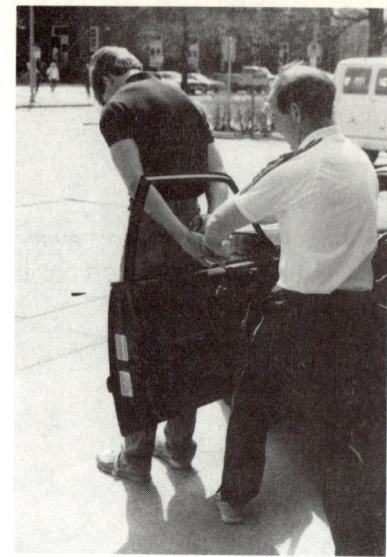

Figure 14-61

Vehicle Hood

The subject of a high risk traffic stop is directed to walk backward toward the squad car until the backs of his knees make contact with the front bumper of the car. (See Fig. 14-61)

Figure 14-62

The subject is then verbally positioned to either the right or left side of the front of the vehicle, dependent upon the selected handcuffing officer. (See Fig. 14-62)

Figure 14-63

The selected officer now advises the subject to lean to the rear, extending his arms rearward, palms out, thumbs up. (See Fig. 14-63)

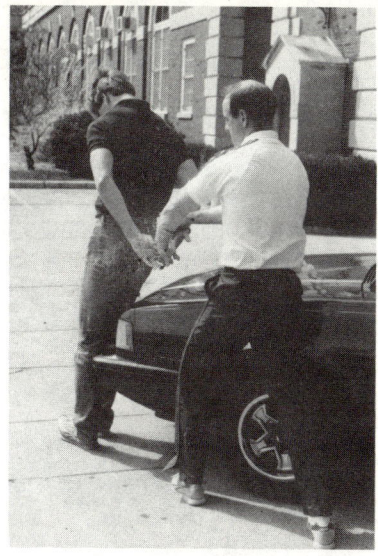

Figure 14-64

The officer now moves quickly from his position behind the car door to the protected position near the fender area of the vehicle and initiates the handcuffing procedure. (See Fig. 14-64)

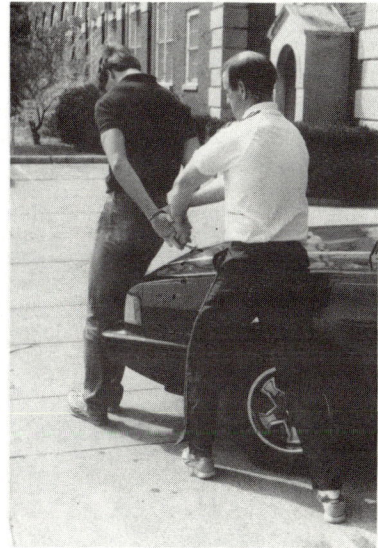

Figure 14-65

Furniture (Table)

The subject is verbally moved to the edge of the table until his rear thighs make contact. (See Fig. 14-65)

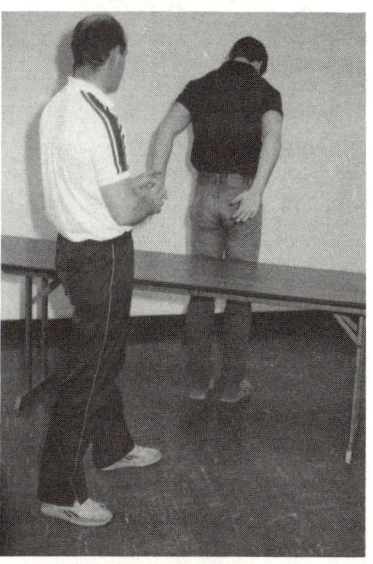

Figure 14-66

He is then ordered to lean to the rear, extending his arms rearward, palms out, thumbs up. (See Fig. 14-66)

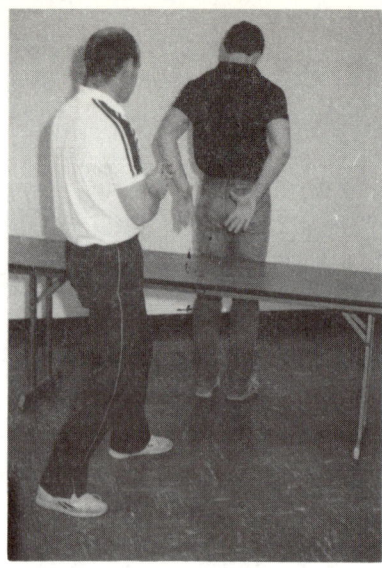

Figure 14-67

The officer now initiates the handcuffing procedure. (See Fig. 14-67)

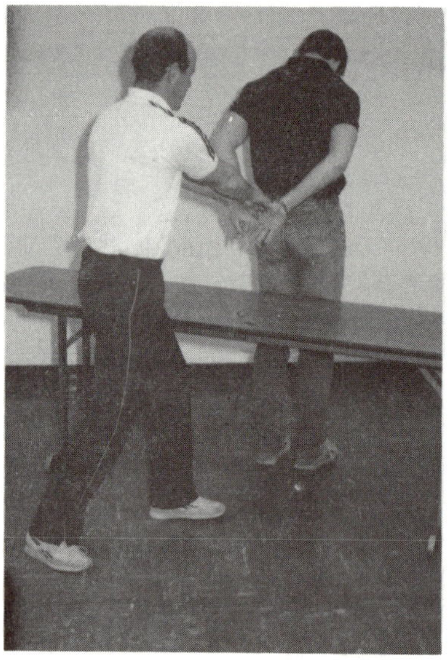

Figure 14-68

Flexible Restraints

Under emergency situations or for specialized handcuffing activities (mass arrests, etc.) flexible handcuffs are acceptable. It is recommended that each officer carry at least two flexible restraints for potential supplemental use.

The same basic rules of application position also apply to the use of flexible restraints as with conventional handcuffs, however, these restraints must first be prepared prior to being applied.

First, hold the restraint with both hands near each end. Thread the narrow, tapered end into the protruding angular side of the head creating a large loop.

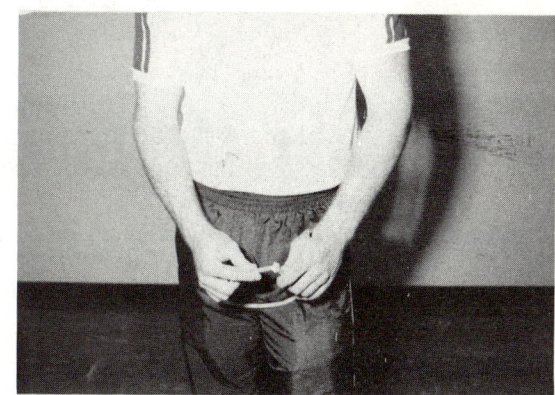

Figure 14-69

Next, interlace a second restraint through the first and create another large loop inter-looped with the first.

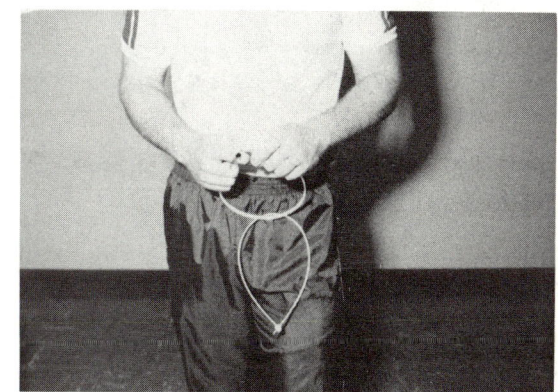

Figure 14-70

A third loop can be inter-looped if desired for more free arm or leg movement after application.

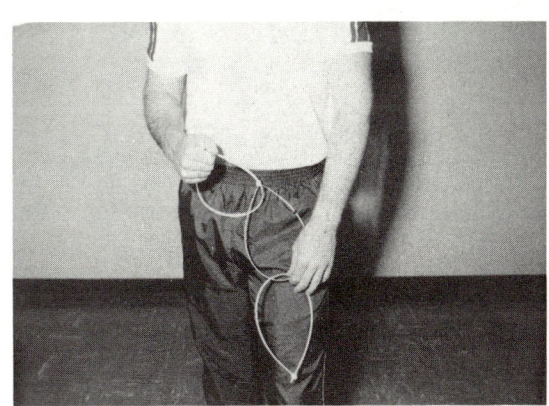

Figure 14-71

175

The subject's strong hand is moved to his back, the restraint is looped over the hand and into position, while the officer's strong hand secures the hand in the finger control grip.

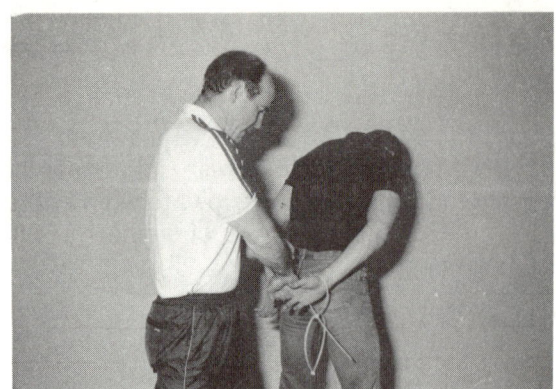

Figure 14-72

The restraint is tightened by the officer's weak hand.

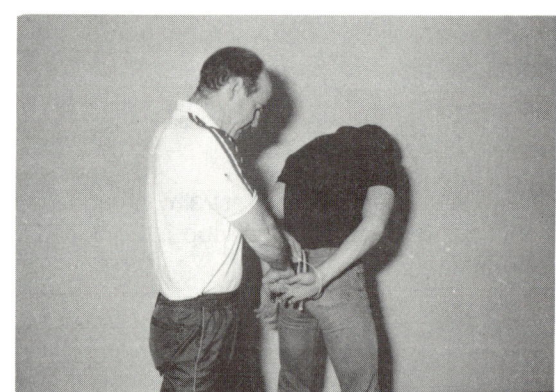

Figure 14-73

The officer continues to secure the strong hand with the finger control grip until the left hand is moved to the back of the subject. The second restraint is looped over the wrist into position.

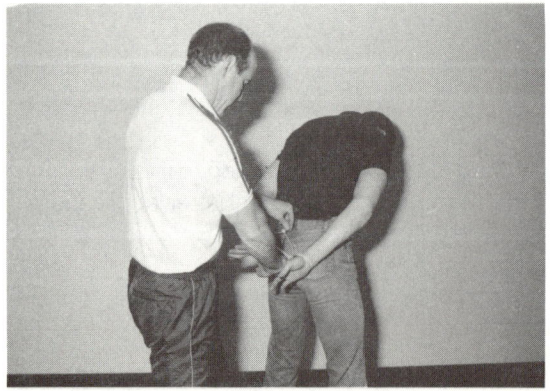

Figure 14-74

Simultaneously the officer secures the remaining hand via his weak hand application of the finger control grip. The second restraint is now tightened by the officer's right hand.

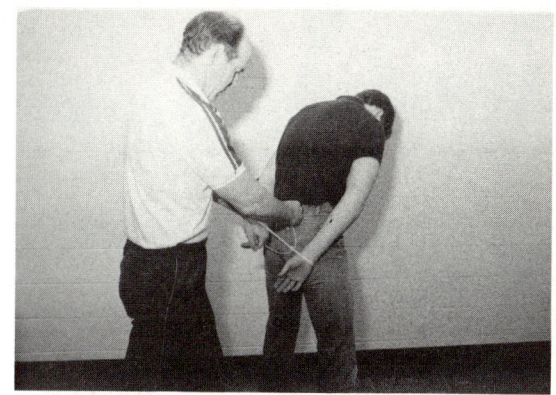

Figure 14-75

It should be noted that several positive attributes can be applied to flexible restraints in addition to emergency or supplement wrist handcuffing.

They can be used to restrain arms together above the elbows,

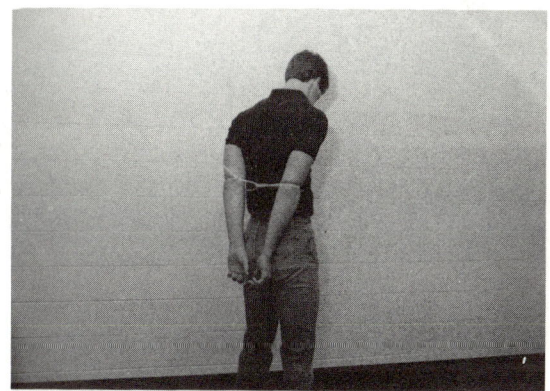

Figure 14-76

restraining the ankles,

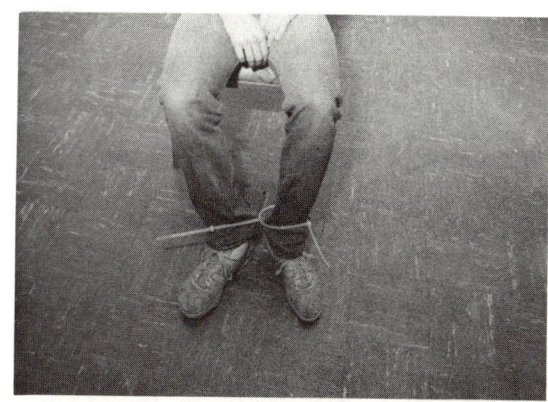

Figure 14-77

177

or restraining the handcuffs to a subject's belt, etc.,

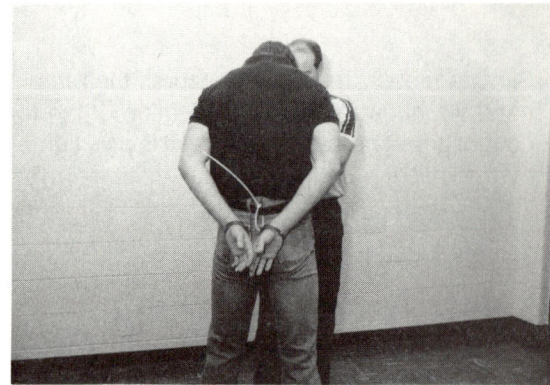

Figure 14-78

Additionally, they can be used to overcuff a subject whose handcuffs must be temporarily removed for loosing purposes.

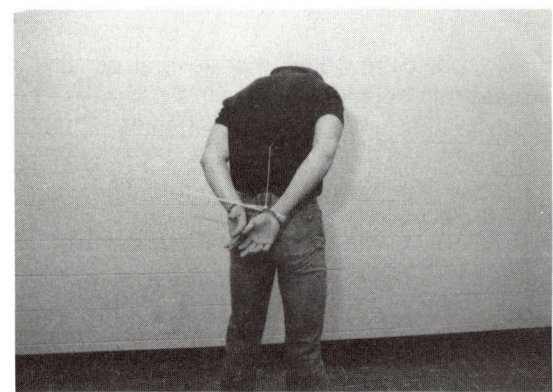

Figure 14-79

Certain negative aspects of flexible restraints must also be noted.

These handcuffs are strictly a temporary restraining device. They can be cut with string or other sawing mechanisms or melted through flame or heating device.

They require the officer to possess a cutting device for most efficient removal.

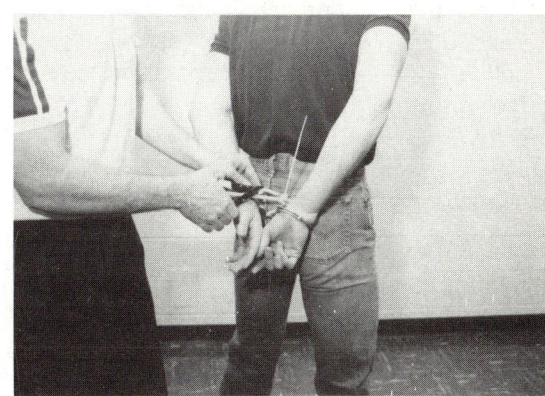

Figure 14-80

Finally, the officer must remain ever vigilant to the occurrence that the subject can pull on the flexible restraints tightening them on the wrist. This may be an attempt to suffer injury or to lure the officer into their removal to effect an escape.

Front Handcuffing

Although the risks are inherent with handcuffing the subject in front, certain activities may mandate such a procedure; court process, document signatures, etc.

In order to facilitate this procedure in a somewhat more safe environment, the following technique with two variations is recommended.

After a proper search has been conducted and the officer is confident that the subject is non-violent, he is ordered to assume a forward kneeling position on a sturdy chair. The subject is told to fully extend his arms, palm down, making a fist with each hand, holding them together.

The officer approaches the subject with the handcuffs in the front, ready position. Here, both cuffs are firmly grasped at the pivot, locks ports facing toward the officer.

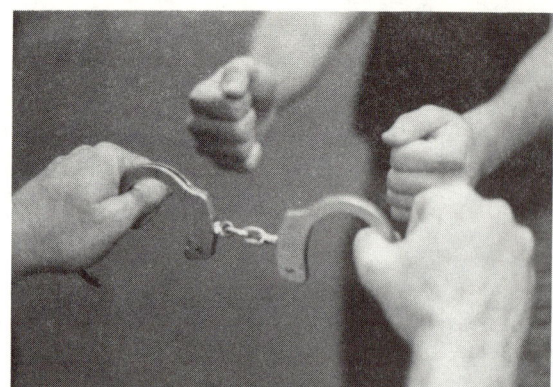

Figure 14-81

The officer now quickly applies the handcuffs over the wrists and double locks them for security.

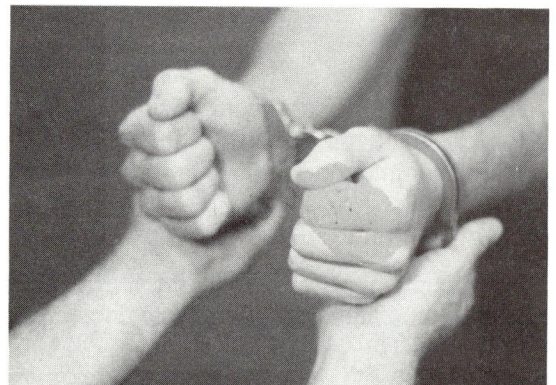

Figure 14-82

The second variation is initiated by ordering the subject to kneel on the floor back turned to the officer.

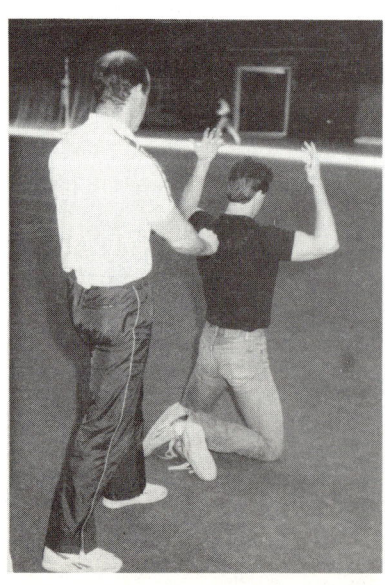

Figure 14-83

The subject sits as far back as possible on his legs to aid in debilitation.

Figure 14-84

The subject raises his hands over his head, holding his fists approximately 2" apart with the thumb back toward the officer.

Figure 14-85

The officer applies the handcuffs from a readied position and double locks the handcuffs.

Handcuffing Resistance Controls

During the handcuffing sequence the subject may attempt to resist the officer's efforts for either escape or assaultive purposes. In such cases the officer must counter the resistance with techniques that correspond to the positioning of the officer throughout the entire handcuffing process. The following techniques are presented as adaptive counter techniques to this type of resistance.

Rear Deployment Armlock

The officer has gained control of the subject's strong hand and arm. The subject rejects the control effort and attempts to break away.

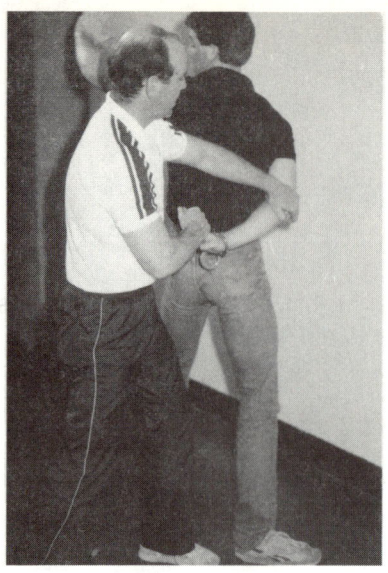

Figure 14-86

The officer increases the tension and upward pressure on the subject's fingers as he pushes the arm to the upper area of the back. (See Fig. 14-86)

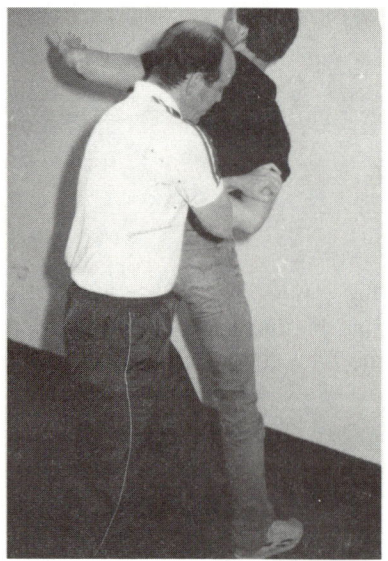

Figure 14-87

The officer now aggressively places his free hand between the subject's forearm, back, and strong shoulder initiating the armlock position. (See Fig. 14-87)

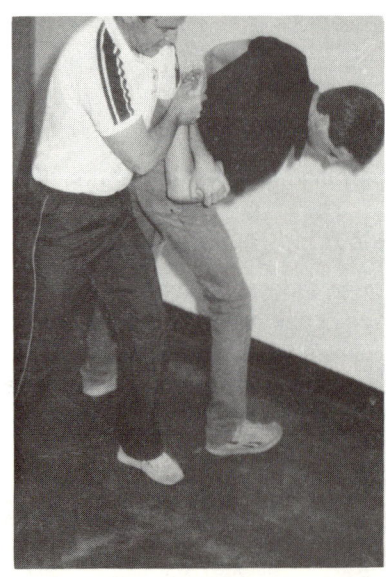

Figure 14-88

The officer rapidly pivots his body toward the rear and begins to spin the subject in a small, powerful, circular motion toward the ground. (See Fig. 14-88)

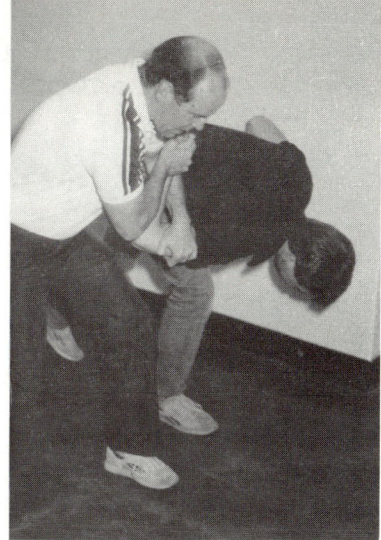

Figure 14-89

The officer concludes the counter move once the subject has been placed in a prone position, compliance has become established, and the prone handcuffing technique is readied. (See Fig. 14-89)

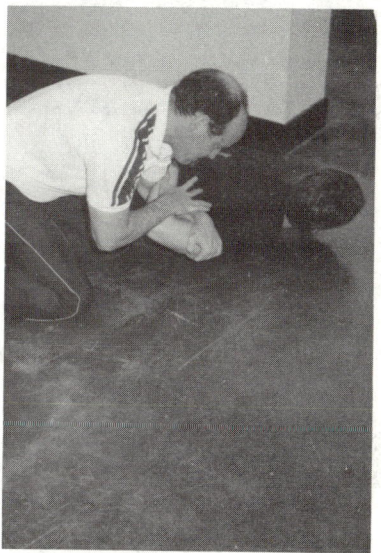

Figure 14-90

Handcuff Pulldown Technique

The subject initiates his resistance once the first handcuff is applied. The officer aggressively jerks the free handcuff, chain, and main frame of the applied handcuff toward the ground.

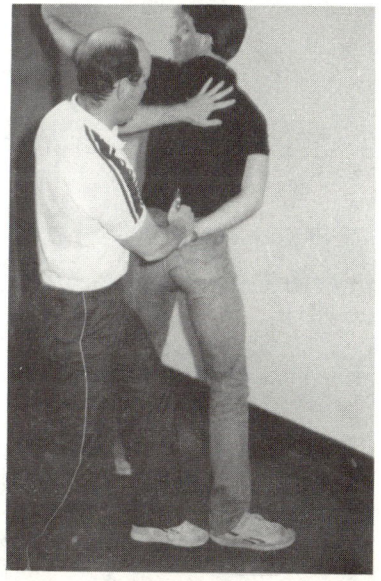

Figure 14-91

Simultaneous to this movement the officer initiates either a shin kick to the common peroneal area of the most available subject knee or a heel kick to the most available tibial nerve area. (See Fig. 14-91, Fig. 14-92)

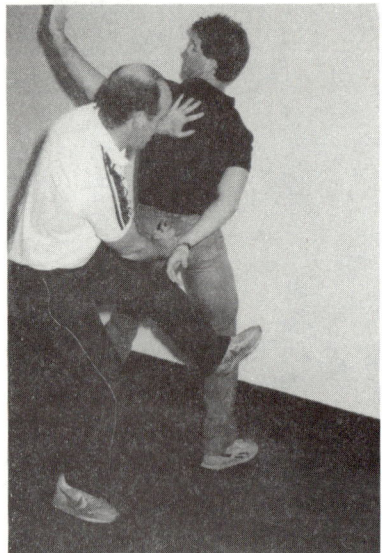

Figure 14-92

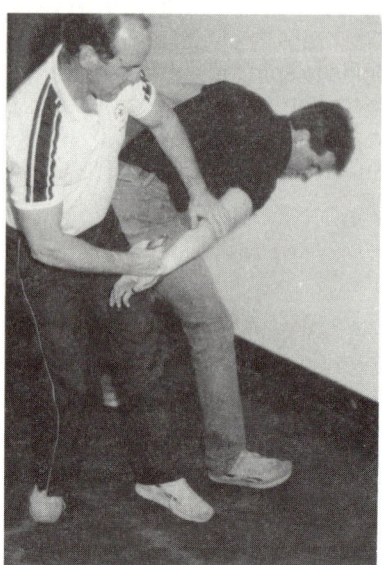

Figure 14-93

These pain compliance and debilitating techniques should de-stabilize the individual for the officer to now commence an appropriate takedown for the subject's placement in a prone position. (See Fig. 14-93)

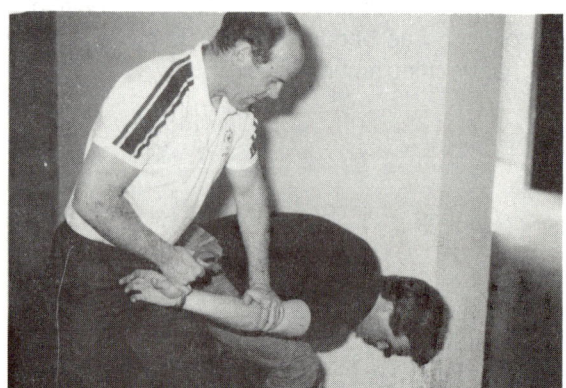

Figure 14-94

Once the subject is secured in this prone position and compliant control is established, handcuffing can begin.

HANDCUFFING APPLICATION EXTREMES

An arrested subject may create restraint difficulties if their wrists are either too large or **too small** for normal handcuff application. In those rare cases where the wrists are so large that normal handcuffs cannot be properly secured, flexible restraints present a viable option. Application can be instituted with either the two or three loop method presented earlier in the text.

If on the other hand the subject's wrists are so small and normal handcuffs are limited in their minimum circumference a modification can be utilized. The handcuffs can be "laced" together decreasing the effective diameter of the handcuffs assuring a more effective restraint.

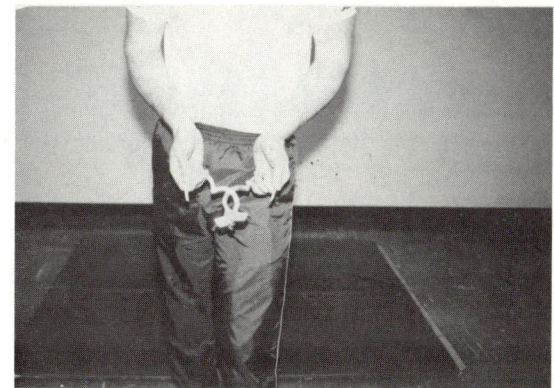

Figure 14-95

One handcuff is simply "laced" inside the other, held in both hands by the officer and readied for either front or rear handcuff application.

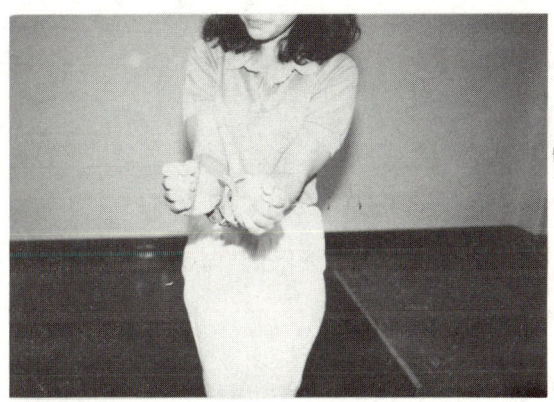

Figure 14-96

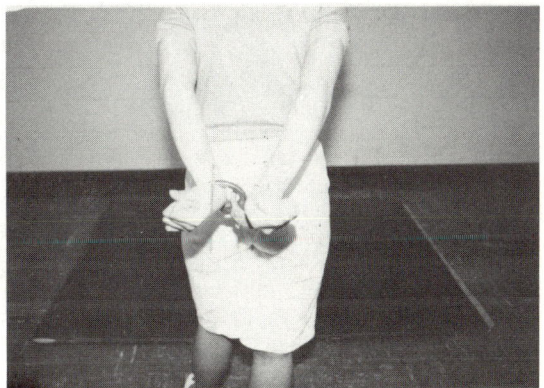

Figure 14-97

Multiple Subject Handcuffing

Two Subjects/One Set of Handcuffs

Once searched the subjects are positioned with their back toward the officer in a debilitated posture either standing or against a wall. (See Fig. 14-97)

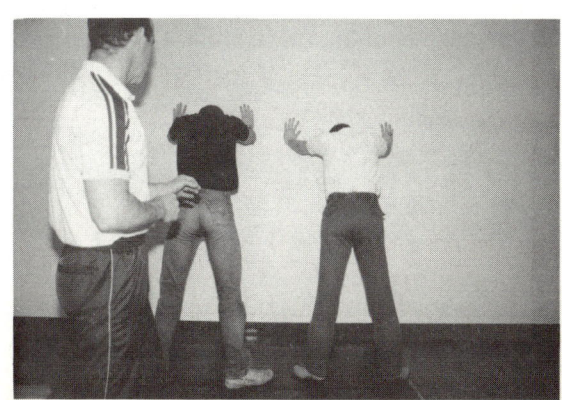

Figure 14-98

185

Each subject is advised to place their right hand to the rear with the thumb up and the palm away. (See Fig. 14-98)

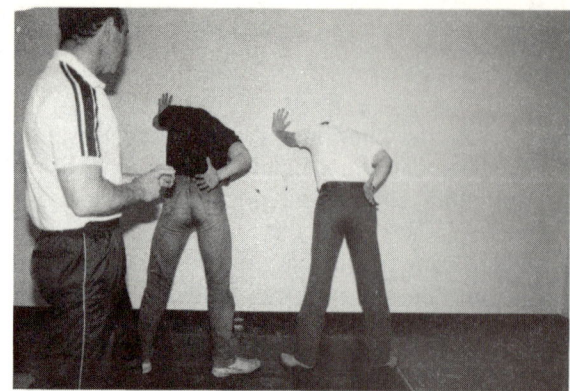

Figure 14-99

The officer initiates control of the subject on the far right and applies the handcuff. (See Fig. 14-99)

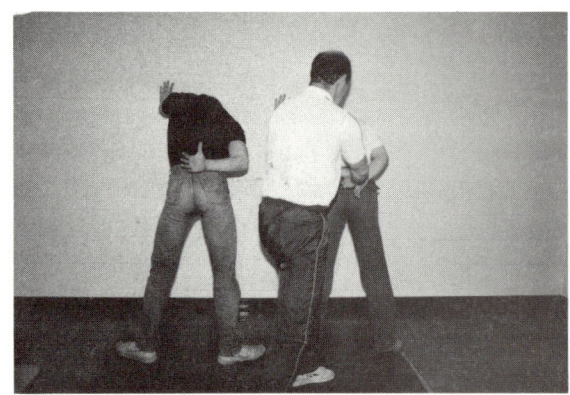

Figure 14-100

He now moves the restrained arm behind the subject in control until it is in position to allow the remaining handcuff to be placed on the remaining subject's right hand. (See Fig. 14-100)

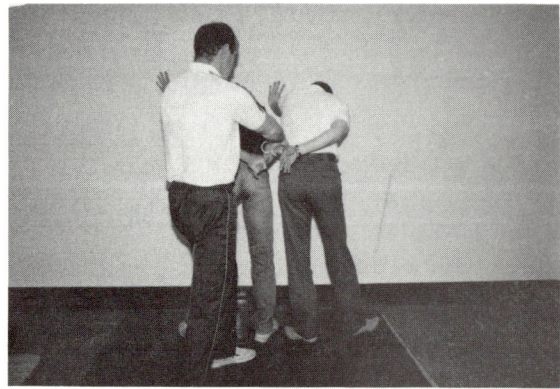

Figure 14-101

In this manner the two subjects will experience debilitation in mobility until a more adequate handcuffing approach can be individually provided. (See Fig. 14-101)

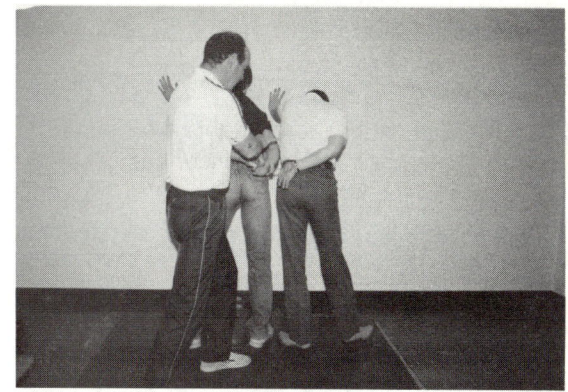

Figure 14-102

Three Subjects/Two sets of Handcuffs

Here, the subjects are placed in a debilitated position on a wall or in a standing position. (See Fig. 14-102)

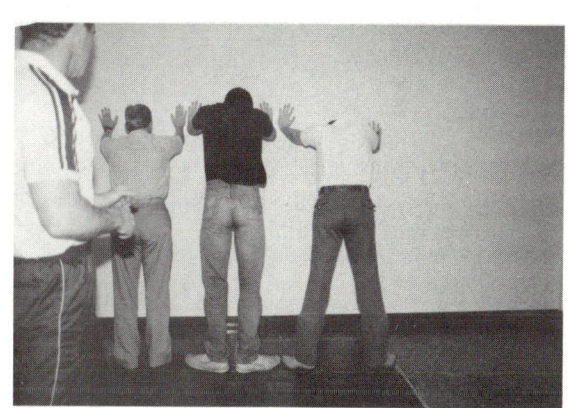

Figure 14-103

The subject in the middle is handcuffed in the traditional manner with increased awareness of the proximity of the other subjects on his right and left. (See Fig. 14-103)

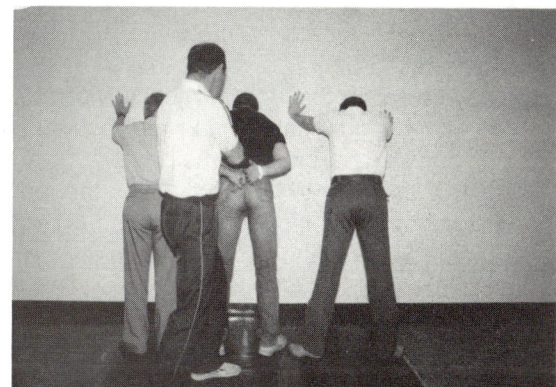

Figure 14-104

The subject on the far right now brings his right arm to the rear until the officer has secured a firm grasp on the fingers and handcuffs the wrist.

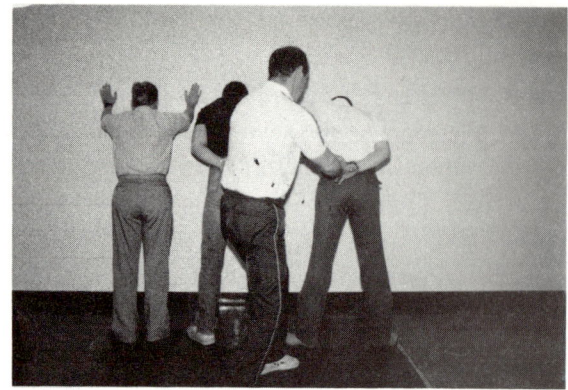

Figure 14-105

The officer now moves the handcuffed hand and arm over to the small of the back of the middle handcuffed individual.

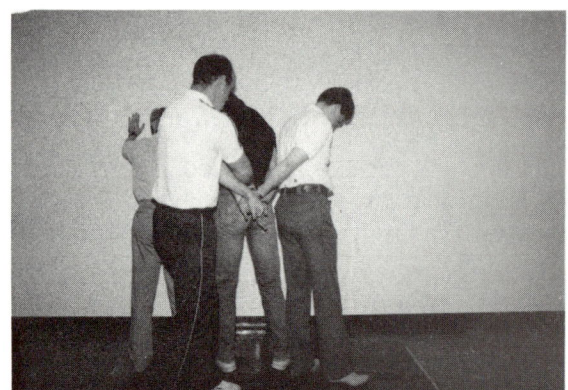

Figure 14-106

The subject on the far left is now advised to bring his right hand to the rear and it is secured by the officer by a firm grasp of the fingers. The officer now laces the free handcuff under the left wrist of the middle subject until it is in contact with the restrained hand and the wrist is handcuffed.

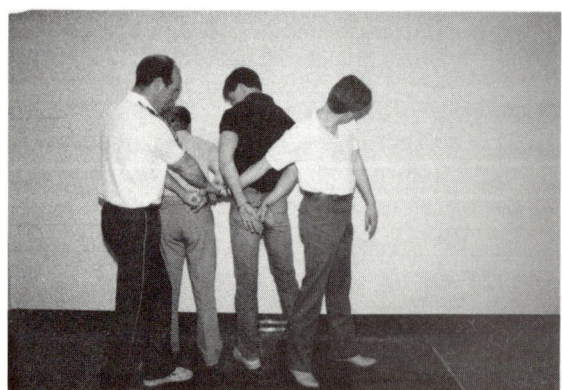

Figure 14-107

Chapter 15

TAKEDOWN/HANDCUFFING COMBINATIONS

As noted previously, our instructional approach is presented within strict confines of time and volume. The time allotment necessitates minimal numbers of techniques presented, with each technique carefully assessed in terms of practicality and clarity.

Volume, in the sense of class size, also specified techniques that are of elementary design so officers can reach skill achievement in mass and in formation to maximize instructor supervision.

Techniques (forms) 1 through 4 capitalize on principles that can easily be related to street encounters. These principles can "equalize" the officer toward his aggressor generally, regardless of that subject's power or strength.

When two forces collide at least two principles are involved. When an object is in motion it has a tendency to remain in motion when acted upon by an outside force. This is commonly called momentum; however, it is correctly called the force of inertia. In other words, if you meet a force "head on", the results of the experience are dependent upon merely the most mass and/or force overpowering the lesser mass and/or force. Inertia plays the vital role in the creation of kinetic energy.

Kinetic energy is the difference in the amount of inertia in these two moving objects upon their inevitable collision. The object containing less inertia (the smaller and/or slower) absorbs the kinetic energy in the collision. This is known as the transfer of kinetic energy. Knowledge of these principles lies at the root of self-defense in the process of mitigating outside forces manifested in punches and kicks that could be aimed for you.

We will examine two methods for diffusing or redirecting outside forces directed toward the officer. Method one involves angular domination and the positive utilization of inertia.

Dynamics of Dominant and Weaker Forces

As depicted in the diagram, you will see that the weaker force was consistently altered to the approximate direction of the dominant force. (See Fig. 15-1) Street applicability will be shown soon as we learn to redirect an adversary's attack.

The second method involves the action of spinning. If a force is directed at a spinning object, the intruding force will be consumed or caught up in the inertia of the spinning object. (See Fig. 15-2) Consequently, if someone charges at you and you spin in the action of defense, you significantly decrease your injury potential. Again, the preceding forms will embody these principles in action terms.

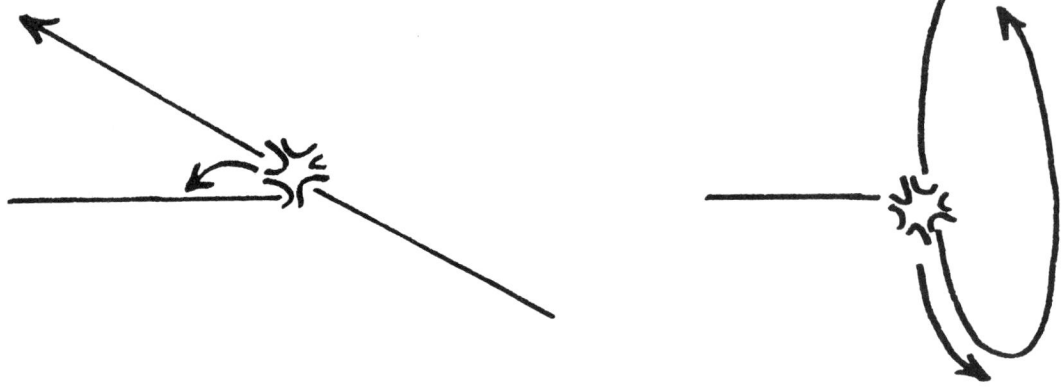

Figure 15-1

Figure 15-2

Form 1

Grasp the subject's wrist with your matching hand (adversary punches with an overhead hammer fist or downward strike) in a modified cross-grip fashion. Place your palm against the back of his wrist, your thumb should hook under the wrist to stabilize the grip. Your thumb should be pointed up his arm while your fingers continue to wrap toward the inside of the arm. (See Fig. 15-3, Fig. 15-4, Fig. 15-5)

Figure 15-3

Figure 15-4

Figure 15-5

Place the palm of your other hand on the underside of the subject's elbow, extending your fingers in the direction of the point of the subject's elbow. (See Fig. 15-6)

Figure 15-6

While stabilizing the elbow area, push the subject's hand toward his chest until you have his forearm and bicep area at a right angle. (See Fig. 15-7) Simultaneously, move the controlled arm so that his elbow is extended directly in front of him while his forearm parallels his chest.

Figure 15-7

To commence the controlling movement, push the subject's elbow directly at his face and pull down on his wrist in a "lever" action. Simultaneously, step, with the leg matching the hand on the subject's elbow, across the front of the subject in a 45° angle from your original position. This action will turn the subject approximately 180°. (See Fig. 15-8, Fig. 15-9, Fig. 15-10, Fig. 15-11, Fig. 15-12 and Fig. 15-13)

Figure 15-8

Figure 15-9

Figure 15-10

Figure 15-11

Figure 15-12

Figure 15-13

Apply downward pressure on the back of the subject's elbow and lift on his wrist while advancing, to flatten the subject onto his stomach. It is essential at this point to tell the person what you expect, i.e., "Go down to your knees, now to your stomach." (See Fig. 15-14, Fig. 15-15, Fig. 15-16, Fig. 15-17 and Fig. 15-18)

Figure 15-14

Figure 15-15

Figure 15-16

Figure 15-17

Figure 15-18

Once the subject is on the ground, keep the controlled arm a 90° minimum angle or more to the rest of his body. Place your knee closest to his affected armpit into the armpit area in a manner reciprocal to the resistance. This 90° or more positional angle must be maintained throughout the control sequence. (See Fig. 15-19)

Figure 15-19

The subject's arm should be pinned to the ground by pushing down on the elbow with the left hand while the officer's right hand pushes down on the suspect's right wrist insuring that the restrained arm is permanently positioned with the palm up. (See Fig. 15-20)

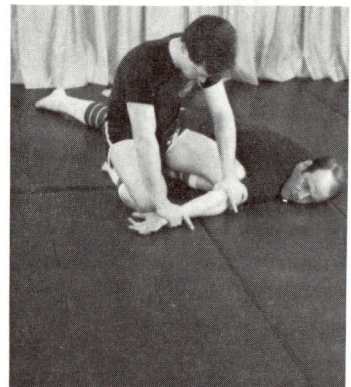

Figure 15-20

At this point, the subject should be handcuffed to maximize control. The officer replaces his right hand control of the subject's wrist with his right knee. (See Fig. 15-21) The officer's right hand moves down to the surface under the subject's palm-up right hand. (See Fig. 15-22)

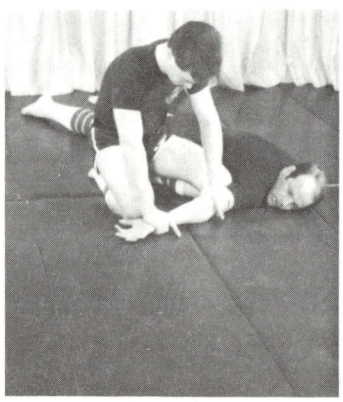

Figure 15-21

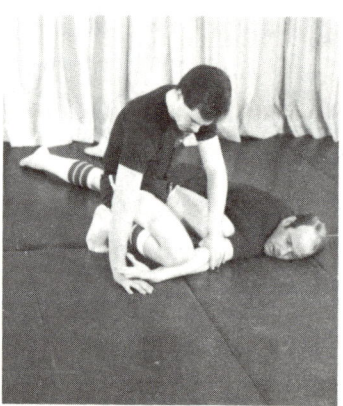

Figure 15-22

The officer places his palm against the backside of the subject's hand, wraps his fingers around the blade edge of his controlled hand, and positions his thumb in the webbing between the subject's thumb and index finger. (See Fig. 15-23)

The officer now applies inward pressure with his palm against the wrist and twisting tension against the wrist and arm by rotating the fingers toward the subject's face at approximately a 45° angle. The subject's entire affected arm with minimal strength or safety commitment by the officer. (See Fig. 15-24)

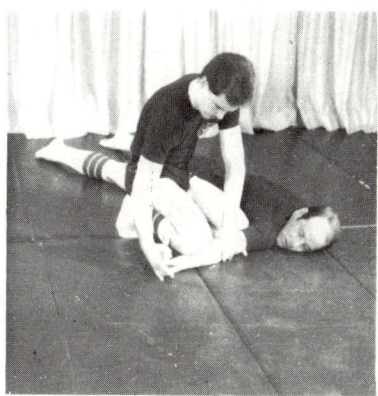

Figure 15-24

Once the subject has been placed in this submission technique, command compliance is more effectively achieved.

Each of the following takedown techniques will continue to conclude with this standardized position of compliant control. In this manner a consistent, reliable technique can be practiced and thus skill levels can be maximized.

Similarly, once the individual is taken to this prone posture the subject will be handcuffed in a standardized manner which will be presented in the latter portion of the chapter.

Form 2

Grasp the subject's wrist in a cross-grip technique placing your palm against the back of his wrist area, while wrapping your fingers around the outside edge of his hand (See Fig. 15-25, Fig. 15-26 and Fig. 15-27)

  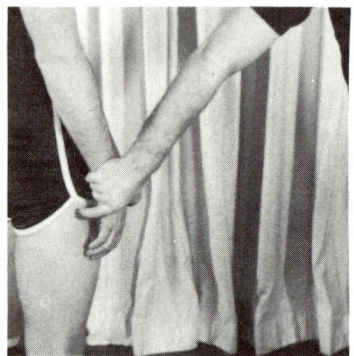

Figure 15-25   Figure 15-26   Figure 15-27

Place your thumb between his thumb and index finger knuckle and rotate the hand with the fingers following in a pattern of a semi-circle, fist downward, continuing around until they are positioned straight up causing the arm to be locked outward. (See Fig. 15-28, Fig. 15-29, Fig. 15-30 and Fig. 15-31)

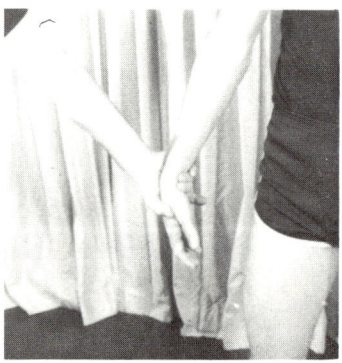 

Figure 15-28   Figure 15-29

Figure 15-30

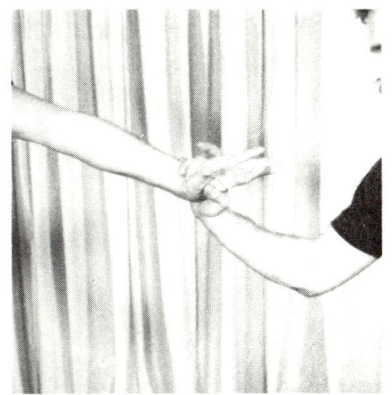

Figure 15-31

The officer steps toward the subject with his weak leg placing the heel of his weak hand on the back of the subject's strong elbow. With the exception of the controlling grip, the position should be virtually identical to that of Form 1 including takedown methodology. (See Fig. 15-32, Fig. 15-33, Fig. 15-34 and Fig. 15-35)

Figure 15-32

Figure 15-33

Figure 15-34

Figure 15-35

Again, *tell* the subject what actions you expect from him so he can comply rather than subject himself to increased joint tension via his resistance.

Once the subject is positioned on his stomach, utilize the control/submission technique in application for Form 1.

Form 3

Grasp the subject's hand and wrist area with the opposite hand of that from the attacker. (See Fig. 15-36)

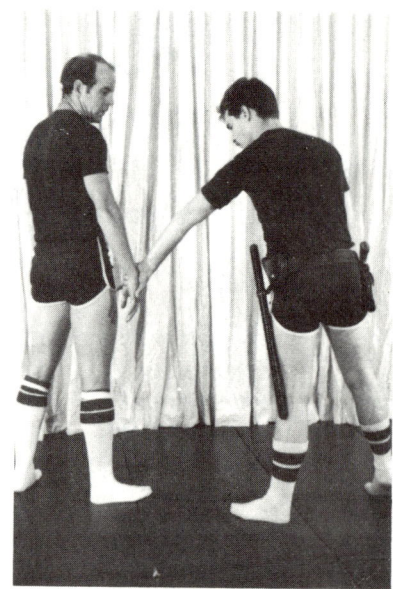

Figure 15-36

Reach across the subject's hand to the thumb side, wrapping your fingers around the thumb side edge placing your thumb in position between his ring finger knuckle and his little finger knuckle. (See Fig. 37)

Figure 15-37

Place the heel of your other hand on top of your thumb exerting additional pressure on your palm. (See Fig. 15-38)

Figure 15-38

Step in the direction of the controlled hand with weak foot, pushing the wrist down and toward the subject or while simultaneously turning the palm outward and away. (See Fig. 15-39 and Fig. 15-40)

Figure 15-39

Figure 15-40

Direct the finger tips on back toward the ground while the subject's forearm and upper arm arc at a right angle. (See Fig. 15-41)

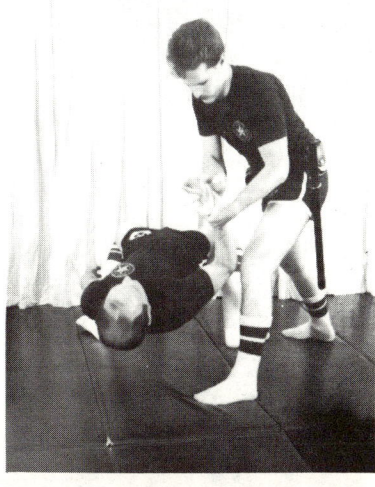

Figure 15-41

Once the person is maneuvered to his back, maintain your grip and walk around the subject's head while turning his wrist and directing the fingertips toward his face and actually pulling him over to his stomach. (See Fig. 15-42, Fig. 15-43, Fig. 15-44, Fig. 15-45, Fig. 15-46 and Fig. 15-47)

Figure 15-42

Figure 15-43

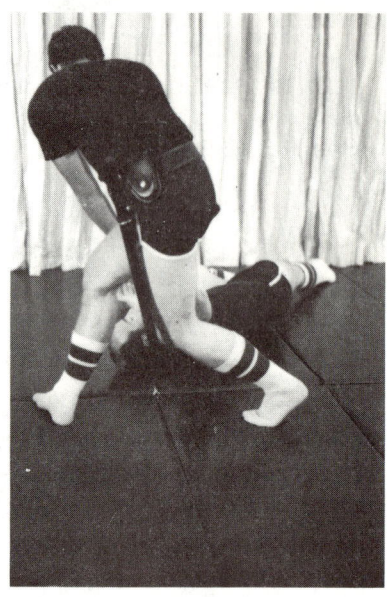
Figure 15-44

Figure 15-45

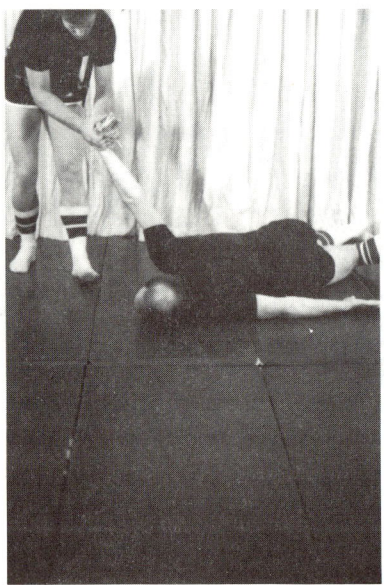
Figure 15-46

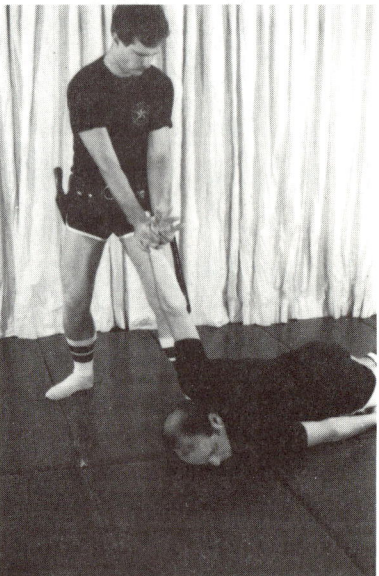
Figure 15-47

Once the individual is tugged into the prone posture, the officer can downwardly collapse his weak knee into position on the tricep/shoulder area, pinning the subject's controlled arm into a secured status. (See Fig. 15-48 and Fig. 15-49)

Figure 15-48

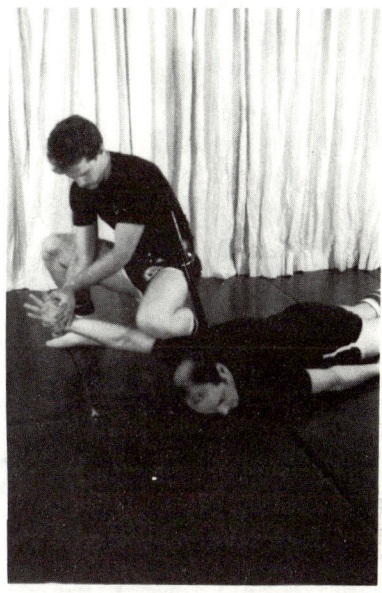

Figure 15-49

At this point, the standard submission/control technique can be effected utilized in Forms 1 and 2.

Form 4

Grasp the subject's fingers in a cross-grip position, placing your palm against the back of his fingers. (See Fig. 15-50 and Fig. 15-51)

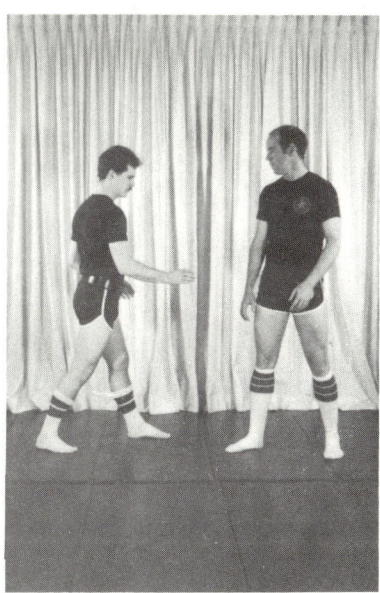

Figure 15-50

Figure 15-51

Step toward the side of the subject that you grabbed. Turn his palm out and push up so that his forearm and bicep area form a right angle, and his arm is parallel to his body. (See Fig. 15-52 and Fig. 15-53)

Figure 15-52

Figure 15-53

Step around so that you face the same direction as the subject. Place the palm of your other hand on the back of his wrist—which should be facing his ribs at this point—wrap your fingers around the edge of his hand—your fingers should touch his palm, if possible—and wrap your thumb around his wrist. This is the primary established grip. (See Fig. 15-54, 15-55, Fig. 15-56 and Fig. 15-57)

Figure 15-54

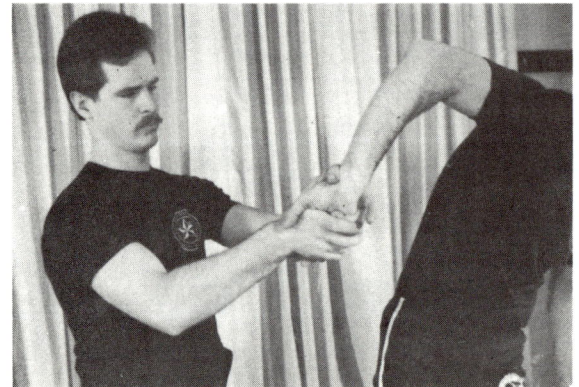

Figure 15-55

202

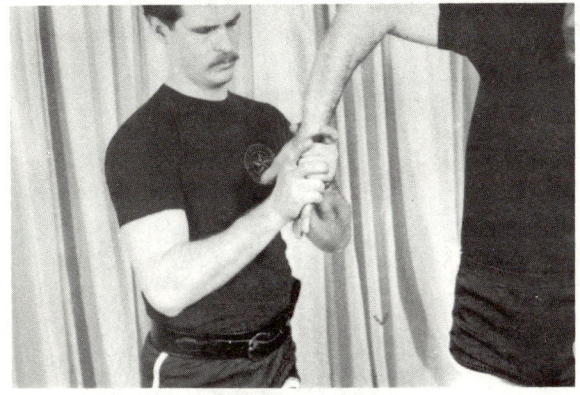
Figure 15-56

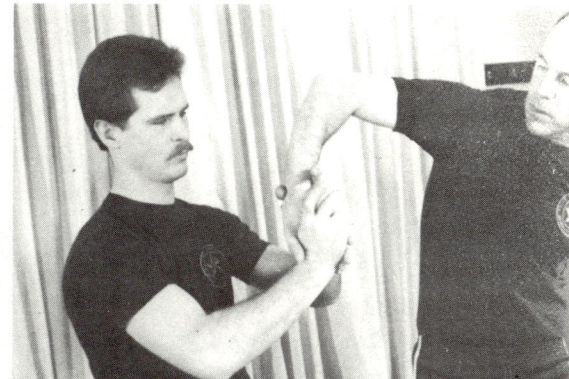
Figure 15-57

The hand that originally picked up the fingers should now slide around to a handshake grip. The subject's fingers, and consequently, his wrist, should be pointing in toward his ribs or straight down—they should not point away from the subject. Submission is achieved by rotating the knuckle of the subject's little finger toward his armpit. You should keep the subject on his toes, if possible. (See Fig. 15-58)

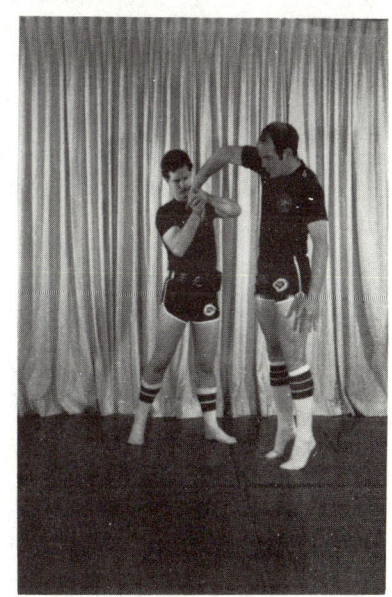
Figure 15-58

To put the subject on his stomach, assume the edge of his hand is a sword, and make a chopping motion parallel to, but inches from his head. This will bend the individual over at his waist. (See Fig. 15-59 and Fig. 15-60)

Figure 15-59

Figure 15-60

Relinquish the *handshake* grip with the right hand and move to the outside of his arm and grab his elbow or interior upper arm. (See Fig. 15-61 and Fig. 15-62)

Figure 15-61

Figure 15-62

The subject should be bent over at the waist and facing you. (See Fig. 15-63 and Fig. 15-64)

Figure 15-63

Figure 15-64

Rotate his palm and elbow toward the ground, pick his arm *up* and place your inside foot beneath his shoulder. (See Fig. 15-65, Fig. 15-66 and Fig. 15-67)

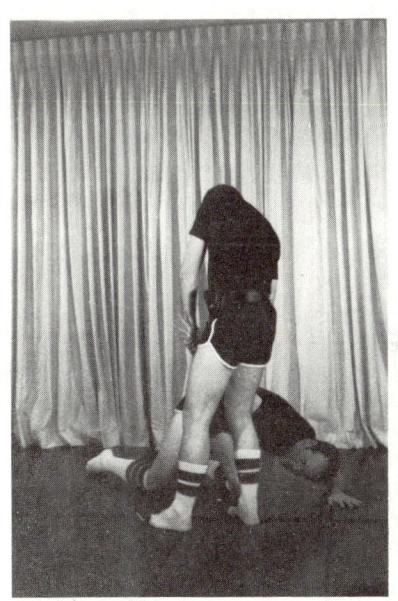

Figure 15-65

Figure 15-66

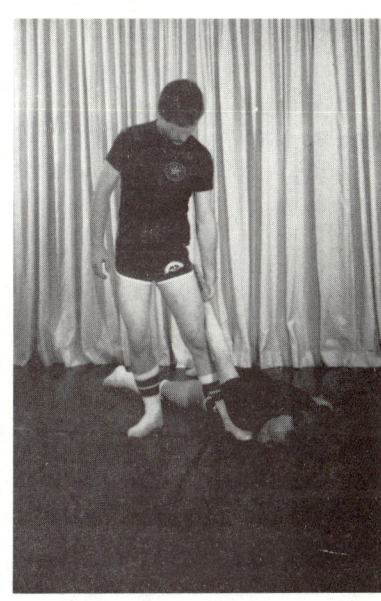

Figure 15-67

Do not lean or bend to the subject, make his arm reach you. Place his palm against the thigh of your left leg. Submission is achieved by rotating the subject's entire arm toward his head.

While tension is continued on the controlled arm the officer can now move to the standard submission/control technique used in Forms 1, 2, and 3. In this case the officer moves his free hand into the Form 2 thumb and index finger web area of the subject. Once this position is secured and a firm grasp is established on the affected hand, tension is increased to maintain control while the officer steps over the arm into a pin out position.

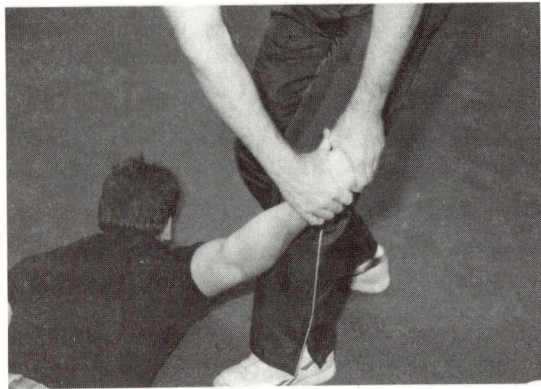
Figure 15-68

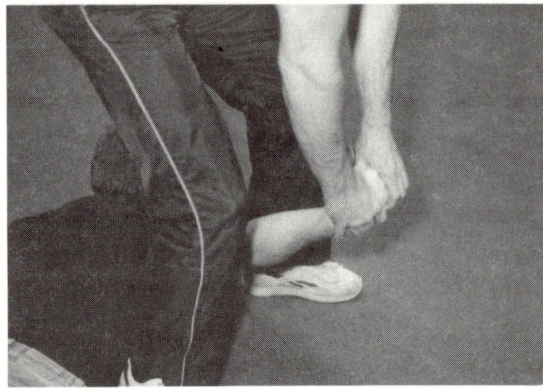
Figure 15-69

The officer continues to maintain tension on the subject's hand, wrist, and forearm via the technique while continuing to apply controlled pressure on the affected arm with his knee. Simultaneously, the officer moves the affected arm into a 90 degree or better position and assumes the standard control status.

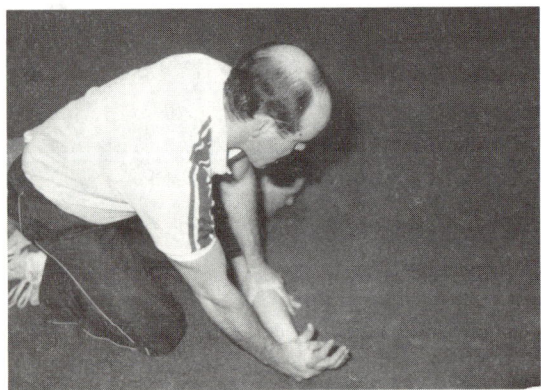
Figure 15-70

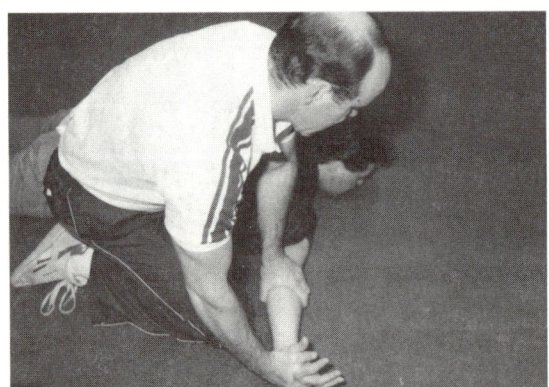
Figure 15-71

PRONE CONTROL HANDCUFFING POSITION

It is suggested that upon the conclusion of any of the takedown techniques utilized (Forms 1, 2, 3, or 4) that the following prone control handcuffing technique be employed.

Once the subject is on the ground, keep the controlled arm a 90° minimum angle or more, to the rest of his body. Place your knee closest to his affected armpit into the armpit area in a manner reciprocal to the resistance. (See Fig. 72)

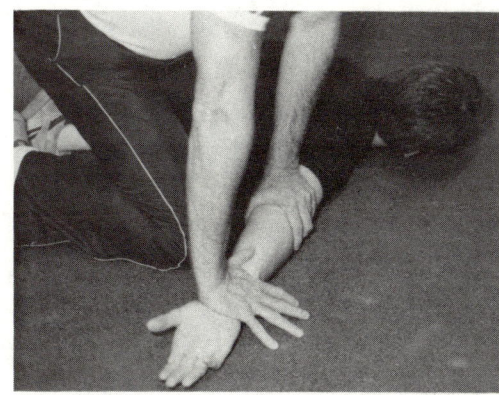

Figure 15-72

The subject's arm should be pinned to the ground by pushing down on the elbow with the left hand while the officer's right hand pushes down on the suspect's right wrist insuring that the restrained arm is permanently positioned with the palm up. (See Fig. 15-73)

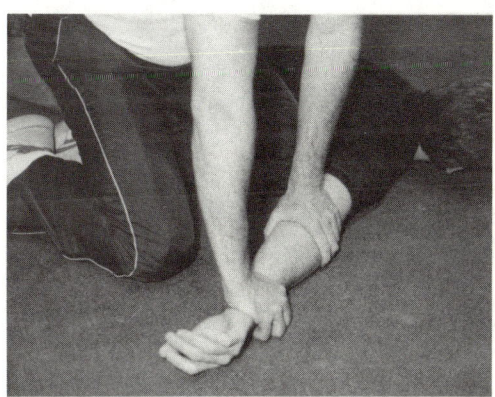

Figure 15-73

At this point, the subject should be handcuffed to maximize control. The officer replaces his right hand control of the subject's wrist with his right knee. (See Fig. 15-74) The officer's right hand moves down to the surface under the subject's palm-up right hand. (See Fig. 75)

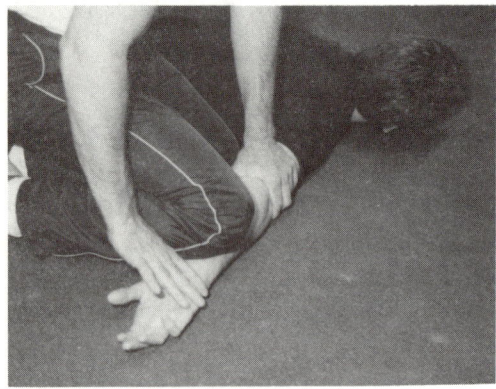

Figure 15-74

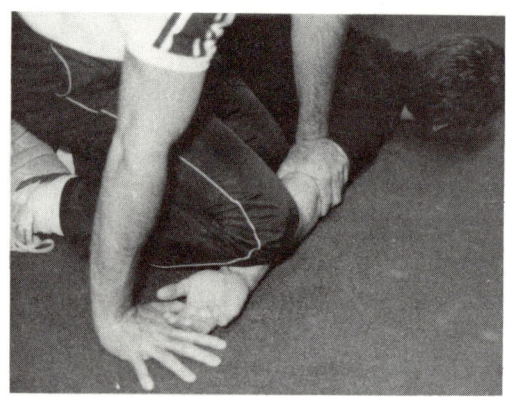

Figure 15-75

The officer places his palm against the backside of the subject's hand, wraps his fingers around the blade edge of his controlled hand, and positions his thumb in the webbing between the subject's thumb and index finger. (See Fig. 15-76)

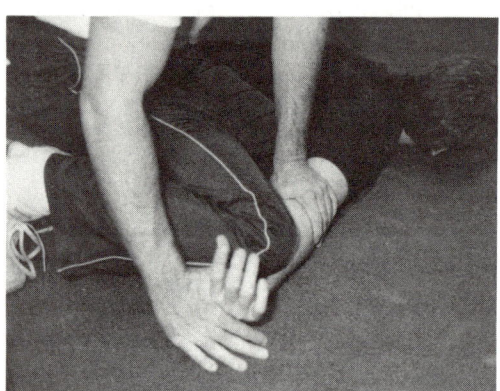

Figure 15-76

208

The officer now applies inward pressure with his palm against the wrist and twisting tension against the wrist and arm by rotating the fingers toward the subject's face at approximately a 45° angle. The submission position causes excruciating pain on the subject's entire affected arm with minimal strength commitment by the officer. (See Fig. 15-77)

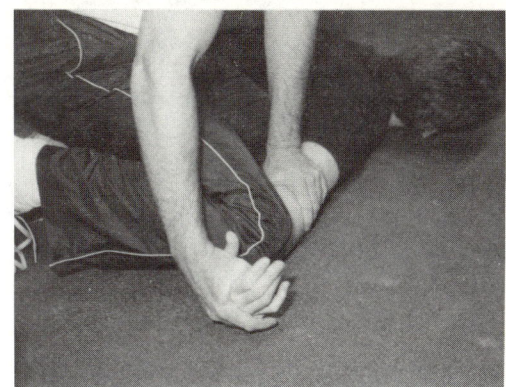

Figure 15-77

Once the subject has been placed in this submission technique, command compliance is more effectively achieved.

The subject is told to place his left hand in the small of his back with the fingers pointed upward. (See Fig. 15-78)

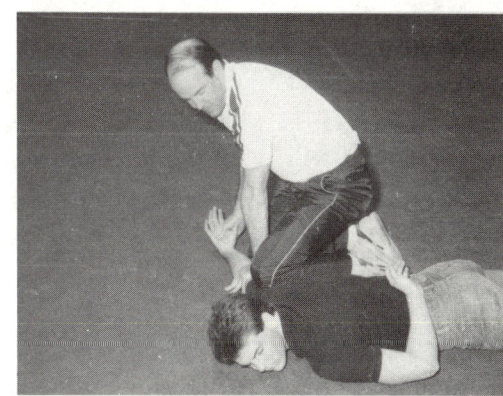

Figure 15-78

At this point the officer replaces his hand control of the subject's armpit with his inside knee. (See Fig. 15-79)

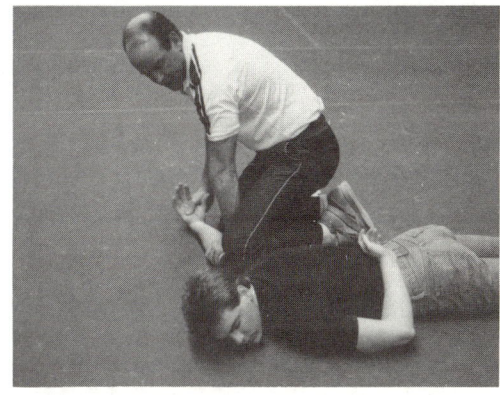

Figure 15-79

The officer now reaches into his handcuff case, removes the handcuffs, initiating the application sequence of the inner handcuff to the subject's left wrist. (See Fig. 15-80 and Fig. 15-81)

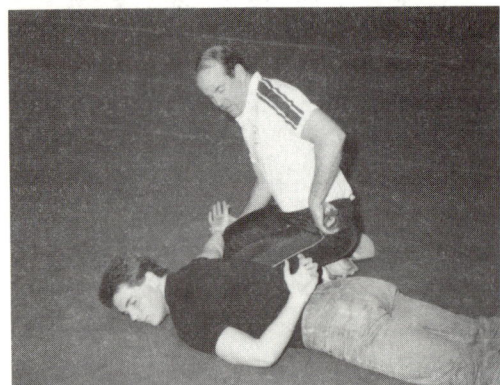

Figure 15-80

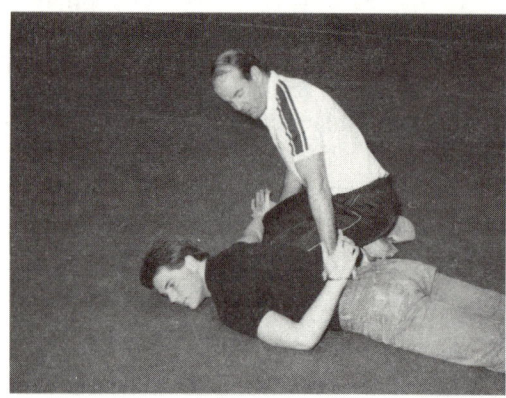

Figure 15-81

The officer's left hand now moves to grasp the closed right handcuff tugging to a maintenance degree against the applied left handcuff. (See Fig. 82)

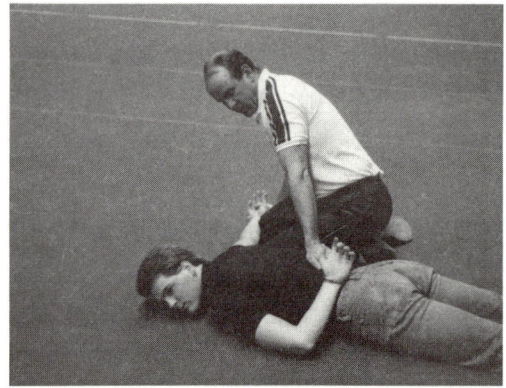

Figure 15-82

At this point primary attention is directed back to the presumed strong hand of the subject. With the subject's right arm in the submission posture, the officer brings his free hand to the back of the subject's elbow. (See Fig. 15-83)

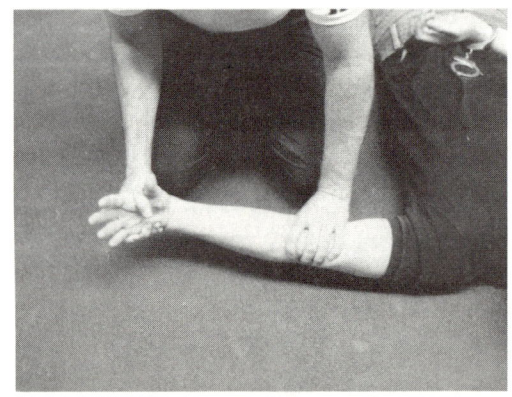

Figure 15-83

The hand placement assists in breaking the elbow into a cradling posture along the officer's forearm. (See Fig. 15-84 and Fig. 15-85)

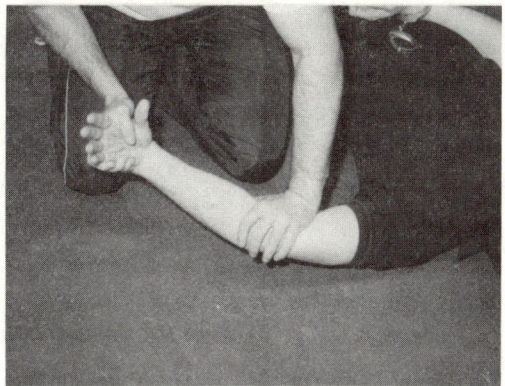

Figure 15-84

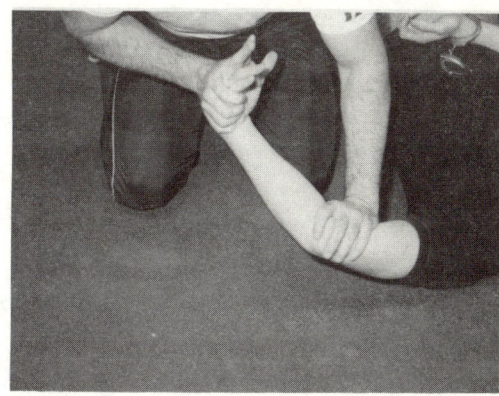

Figure 15-85

The officer positions himself perpendicular to the subject's body continuing to control the cradled arm, eventually floating the grasp into a backfist-wrist control lock. (See Fig. 15-86 and Fig. 15-87)

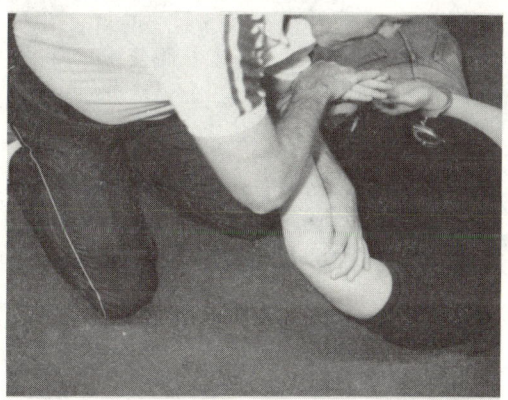

Figure 15-86

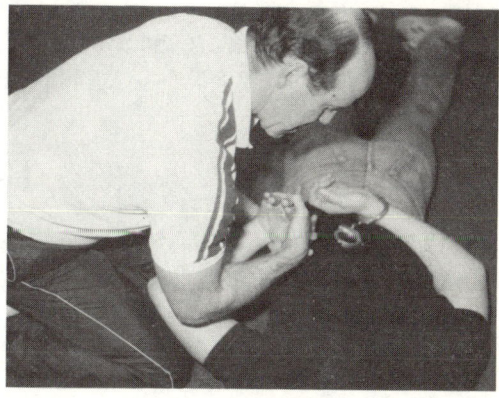

Figure 15-87

The control lock is reinforced when the free hand is placed into the same backfist-wrist configuration. (See Fig. 15-88)

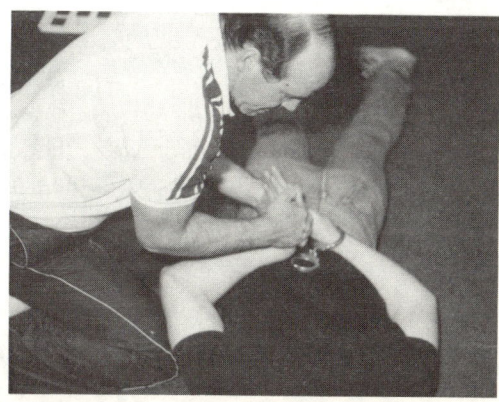

Figure 15-88

The subject's arm is now moved to the small of his back in position for eventual handcuffing. (See Fig. 15-89 and Fig. 15-90)

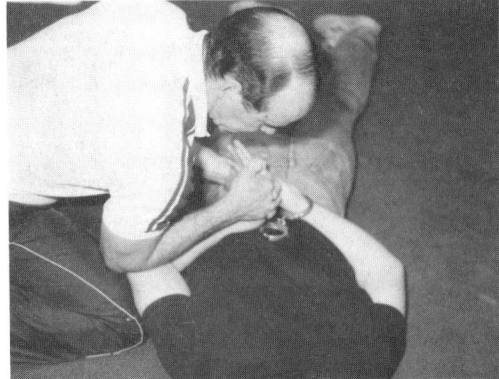

Figure 15-89

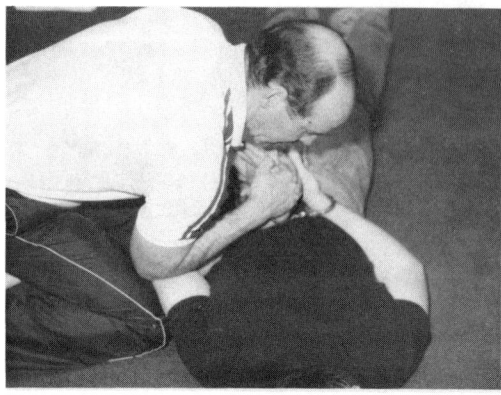

Figure 15-90

The officer's free hand eventually moves the free handcuff on to the subject's wrist. (See Fig. 15-91)

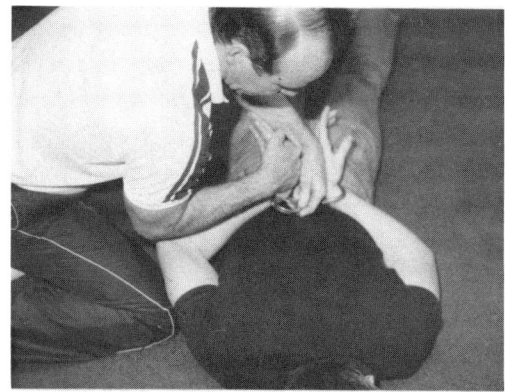

Figure 15-91

HOLD DOWN TECHNIQUES

Circumstances may dictate that a handcuffed subject be held in a prone position for an extended period. The following two techniques are recommended for such purposes, the Head Control and the Leg Pin techniques.

Head Control

Once the subject is handcuffed, the officer moves to the front of the subject controlling the head with his hands. (See Fig. 15-92)

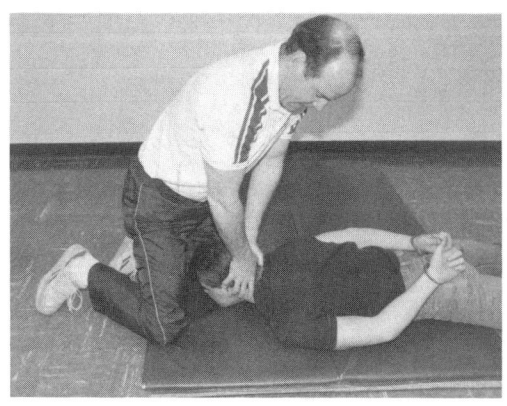

Figure 15-92

He then carefully moves his legs into a position where both thighs are placed along the sides of the subject's head. The officer now applies controlled pressure against the sides of the head, while continuing to secure the head with both hands.

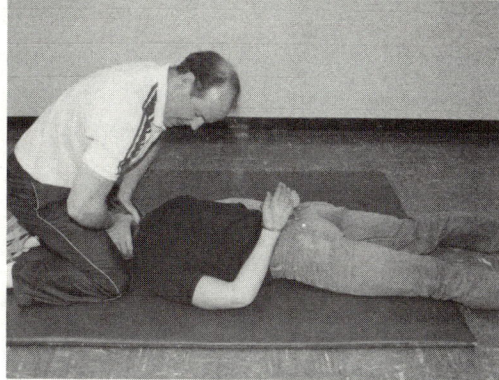
Figure 15-93

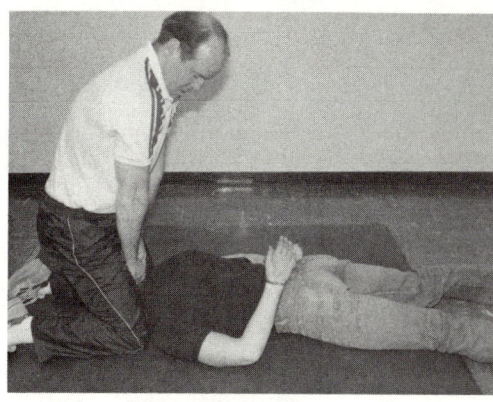
Figure 15-94

If additional control is necessary the officer may utilize three enhancement tactics to assure compliance. He can move to place additional weight onto the via either downward pressure with his hands or actual body mass. He can utilize nerve compression techniques such as the mandibular angle to gain control. Or he can pull the handcuffed hands upward toward the head maximizing pressure on the subject's upper torso.

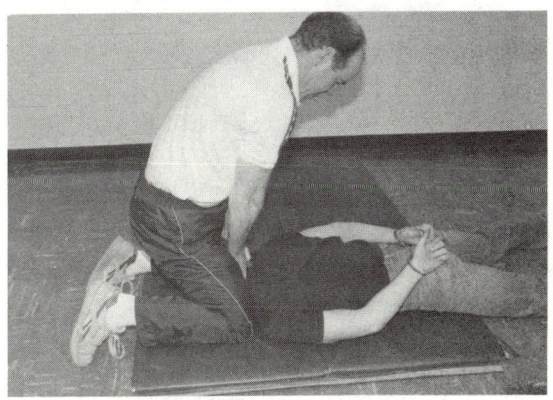
Figure 15-95

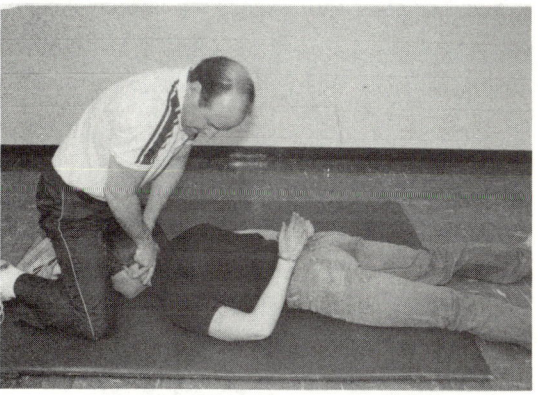
Figure 15-96

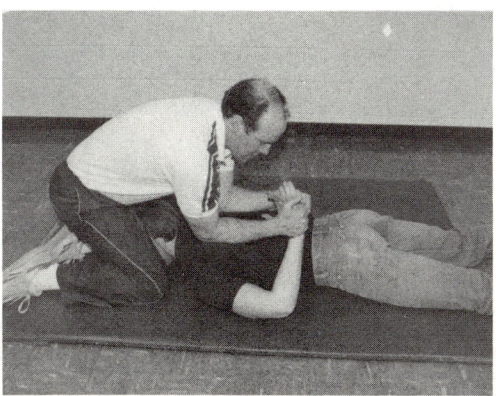
Figure 15-97

Leg Pin

After handcuffs have been applied the officer straddles the subject from the buttocks area.

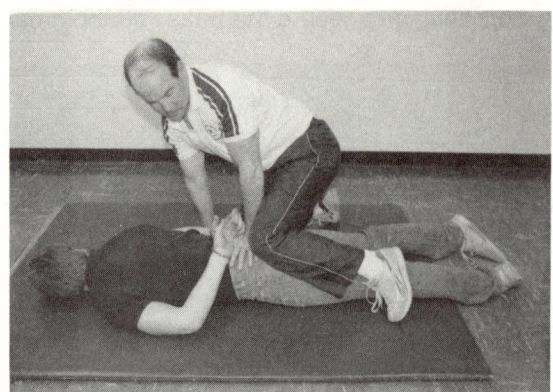

Figure 15-98

The officer begins to compress the subject's legs together with the aid of his hands, arms, and legs. The officer continues to move down the legs until he reaches the calf area where he now allows full body weight to rest on the legs concluding the pinning sequence.

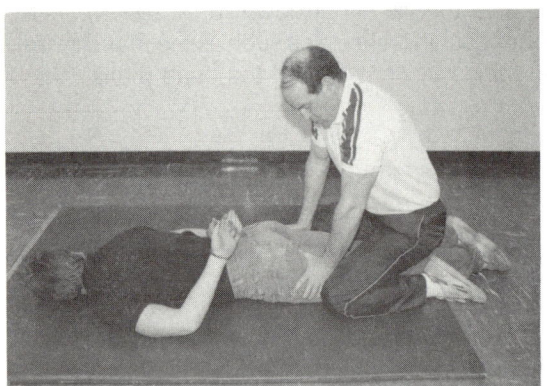

Figure 15-99

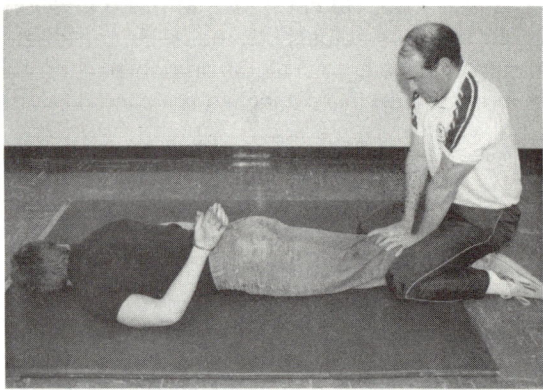

Figure 15-100

If additional control is required the officer can either move further down the subject's legs to the ankle area to maximize the pinning position. This additionally allows the officer access to nerve compression techniques applied to the tibial nerve area of the rear calf.

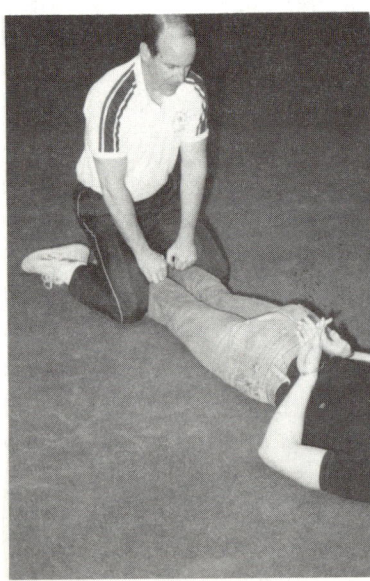

Figure 15-101

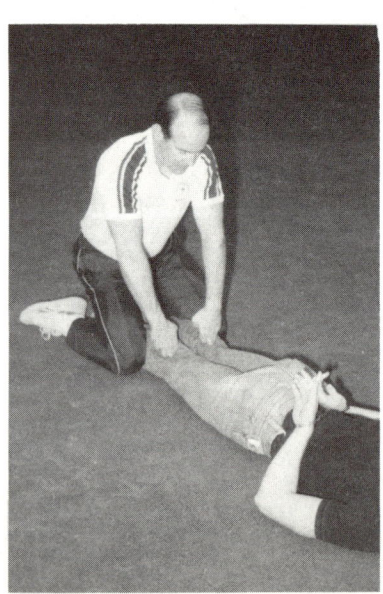

Figure 15-102

Chapter 16

DRIVER REMOVAL TECHNIQUES

With growing frequency, your arrest encounters will occur not in a face to face street contact, but rather as a result of a traffic stop and subsequent arrest. It is not unusual for the arrested subject to seek refuge within his vehicle and attempt to thwart the officer's custodial efforts.

Certainly, your perceptual awareness should be heightened now that the subject has offered resistence to your arrest order. Observe his hands, watch for any unusual movements, and remember he is operating from a position of advantage. Perhaps additional decisive dialogue will convince the suspect to cooperate, but as a last resort, a technique of control may have to be instituted.

We have included three removal techniques, two of which are essentially a reapplication of controlling tactics earlier defined as Form 2 and Form 4. The remaining technique is a time tested method founded upon head/neck control and submission, rather than that of hand and wrist.

It should be noted that the sequencing of the three techniques are based upon the author's shared experience of street effectiveness and their situational adaptability.

Form 4 (Removal Adaptation)

Here, we first open the driver side door of the suspect vehicle. In most cases, the subject will attempt to gain a superior posture by gripping the steering wheel.

In order to secure a controlling grip on the subject's weak hand, a distractive method must be employed. Perhaps three of the best distractions for this situation are the pinching of the subject's left shoulder or neck region with the officer's right hand (See Fig. 16-1), an index finger thrust into the mandibular angle area (See Fig. 16-2), or a nose clamp-thumb to the mandibular angle position. (See Fig. 16-3 and Fig. 16-4)

Figure 16-1

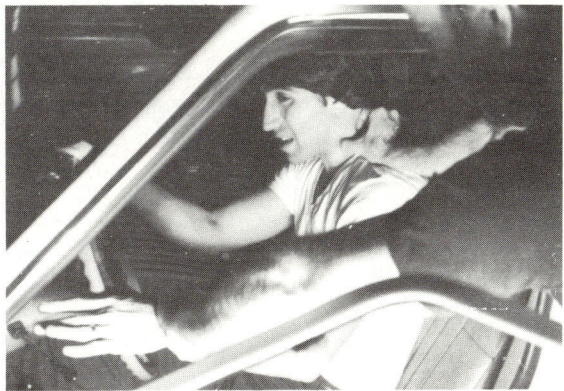
Figure 16-2

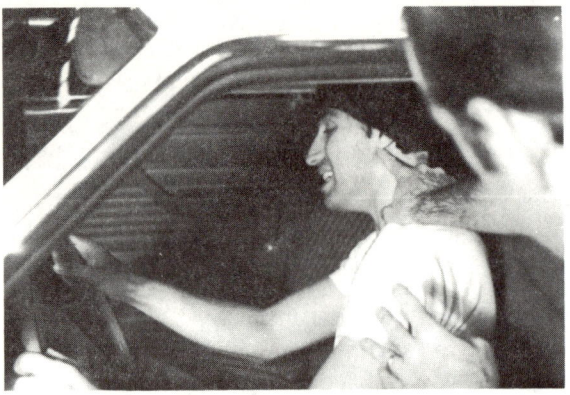
Figure 16-3

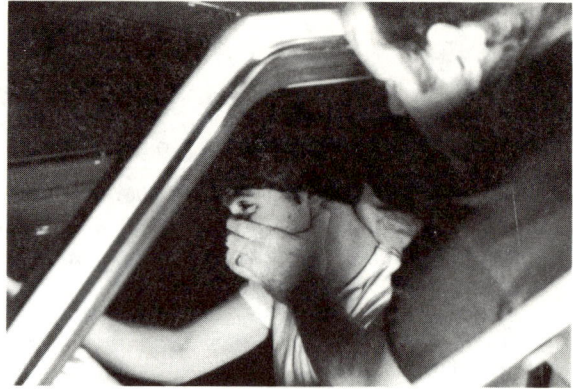
Figure 16-4

The right hand of the officer should be the distraction tool leaving the left hand free for the primary task of subject control. At the point of maximum distraction, pull abruptly on the subject's left hand and wrist. (See Fig. 16-5) Upon breaking the grasp of the subject on the steering wheel, pull the subject's left hand toward the rear of the vehicle. (See Fig. 16-6)

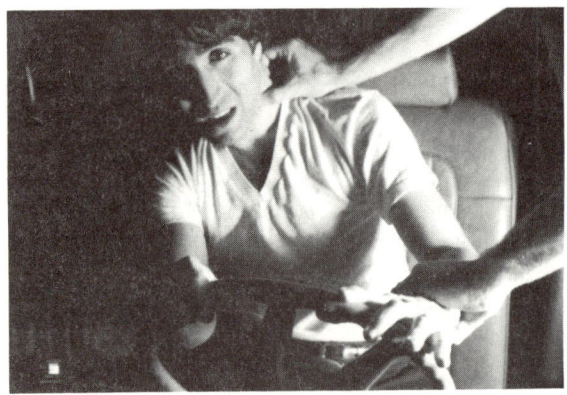
Figure 16-5

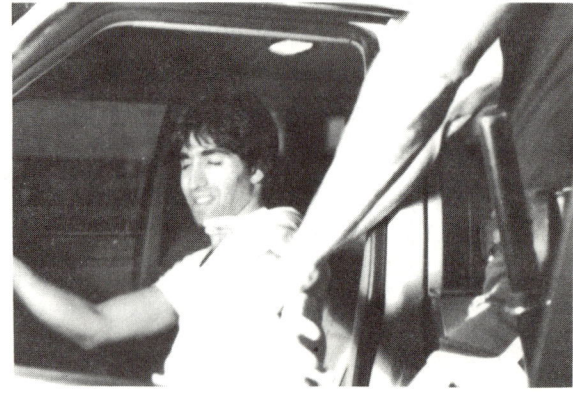
Figure 16-6

Place your right hand under the back of his left hand and hook your fingers over the blade edge and onto the palm (Form 4). (See Fig. 16-7)

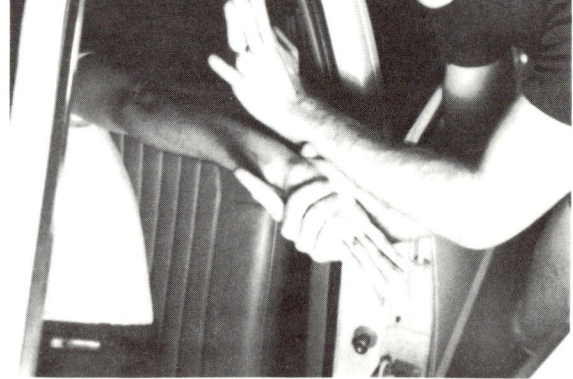
Figure 16-7

216

Assume the wrist twist position by placing your left hand in a handshake position to increase potential tension on the subject's wrist. (See Fig. 16-8)

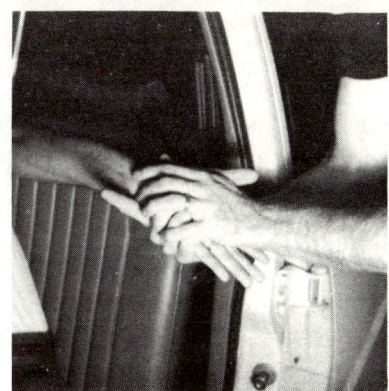

Figure 16-8

Increase the tension to the degree necessary to cause the subject to step out of his vehicle and into an eventual handcuffing posture. (See Fig. 16-9 and Fig. 16-10)

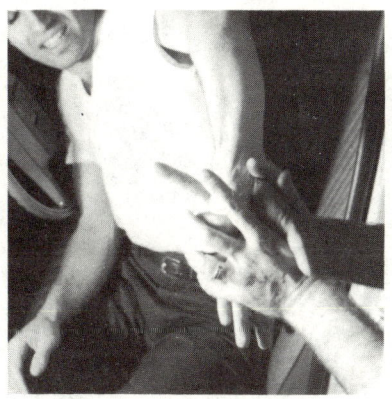

Figure 16-9

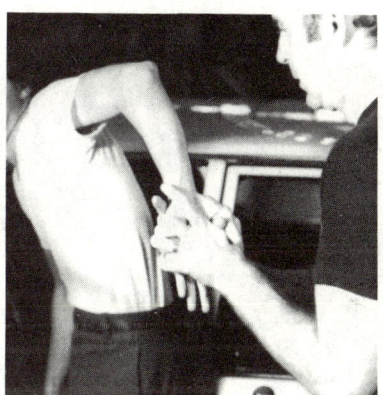

Figure 16-10

Form 2 (Removal Adaptation)

Once you have opened the door of the vehicle and gained a position of advantage next to the driver, initiate an act of distraction to the subject. Such an act could include a finger into the mastoid area, a controlled pull of the hair, etc., conducted with the right hand. (See Fig. 16-11)

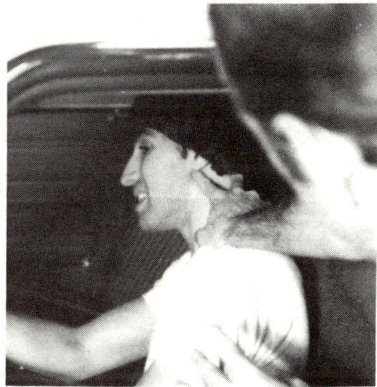

Figure 16-11

Simultaneously, the officer's left hand should grasp the steering wheel of the vehicle and rotate it in a counter-clockwise motion moving the driver's right hand ideally toward the twelve o'clock position on the wheel. (See Fig. 16-12)

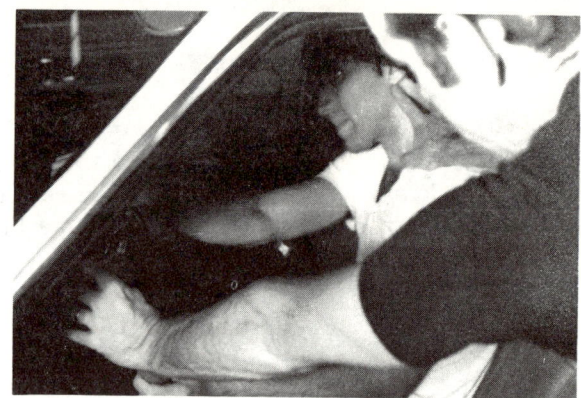

Figure 16-12

(It must be noted that on some late model vehicles, once the engine is off and the ignition switch is turned to the locked position, the steering wheel cannot be rotated. In this case, the preceding control tactic may be more difficult to achieve and should not be attempted.)

From the distracting activity, the officer's right hand should move rapidly down to the subject's right hand and establish a cross-grip grasp on the hand. (See Fig. 16-13 and Fig. 16-14)

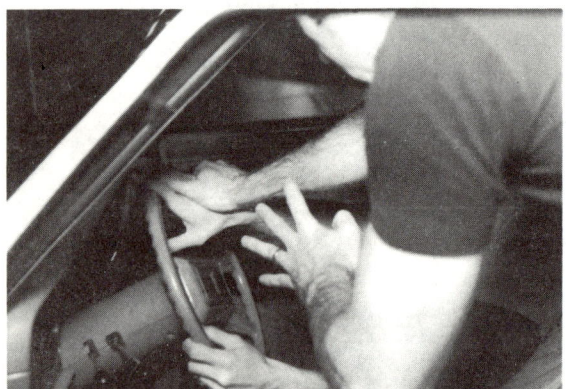

Figure 16-13

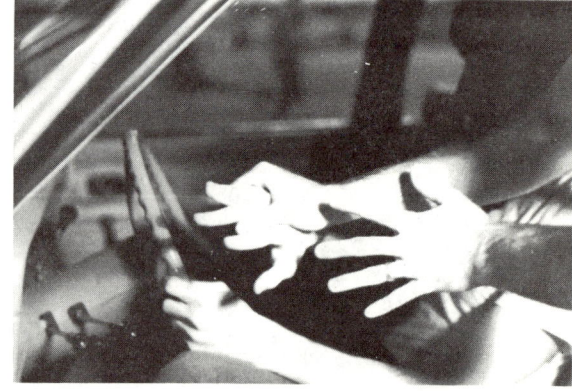

Figure 16-14

The officer's palm should be against the wrist area, while wrapping his fingers around the outside edge of the driver's right hand. (See Fig. 16-15)

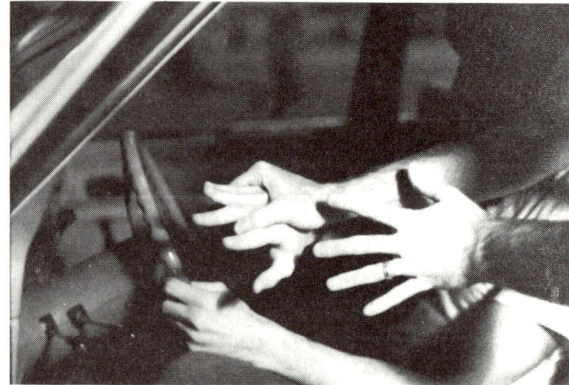

Figure 16-15

The officer's thumb should move between the subject's index and middle finger to peel the hand from the steering wheel, and to rotate the hand in a semi-circle, palm upward. (See Fig. 16-16)

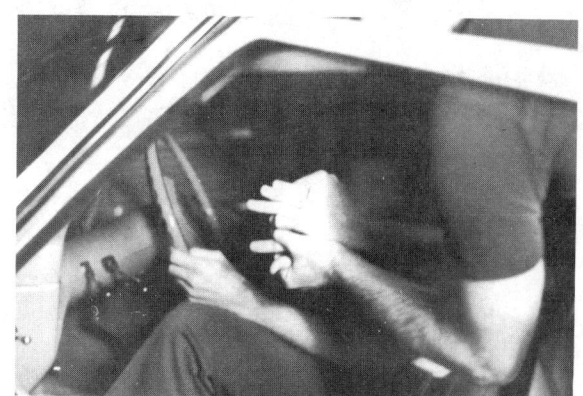

Figure 16-16

The rotating motion can be assisted by the officer moving his free hand to a control position to the back of the subject's elbow. (The officer's right hand thumb is now placed between the index and middle fingers of the driver with the officer's fingers wrapped around the thumb side of the hand.) (See Fig. 16-17)

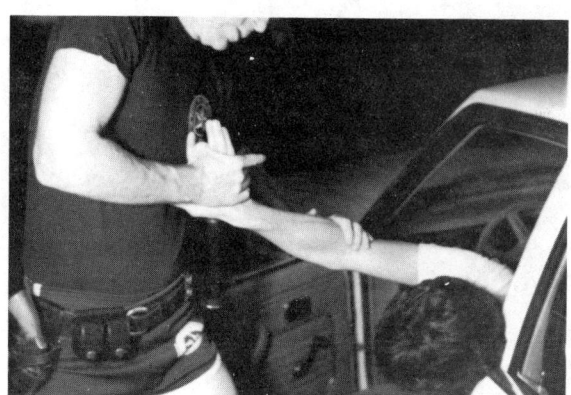

Figure 16-17

At this point, the officer maximizes the pressure against the backside of the subject's right hand which increases correspondingly the tension on the wrist joint. Here the officer should first attempt to achieve verbal compliance to exit the vehicle and begin to submit to the arrest situation. If verbal compliance is not forthwith, the physical route of submission and control is necessitated. The subject is pulled further out of the vehicle while his fingers are laterally pointed toward his head and elbow pressure is applied. Slow tension control promotes the eventual removal activity. (See Fig. 16-18)

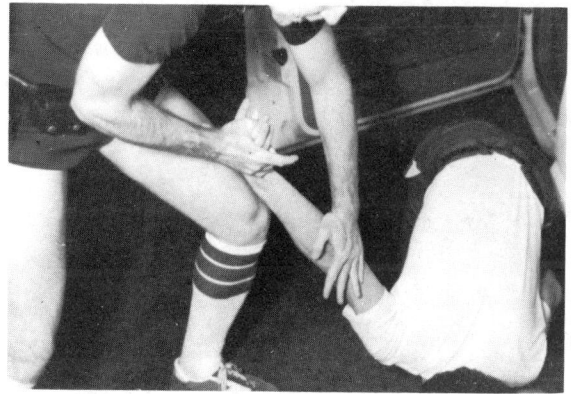

Figure 16-18

Since the subject has demonstrated an uncooperative attitude throughout the encounter, it is suggested that he be removed from the vehicle in a manner that concludes in a prone position ready for handcuff application to afford maximum officer protection.

Head/Neck Twist

Upon opening the driver's door, the officer assumes an alert stance to the side of the subject. Following a verbal request to exit and its noncompliance, the officer places the palm and heel of his left hand cautiously upon the chin of the driver. (See Fig. 16-19)

Figure 16-19

Simultaneously, he pushes the head and face to the right. (See Fig. 16-20)

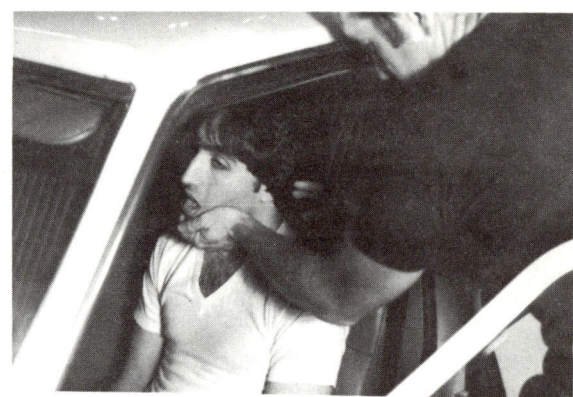

Figure 16-20

Immediate to this position the officer moves his right hand around the rear of the subject's head and assumes position above the subject's right ear/mastoid region. The move allows tension and torque on the driver's neck usually concluding in a suspect's exit. (See Fig. 16-21)

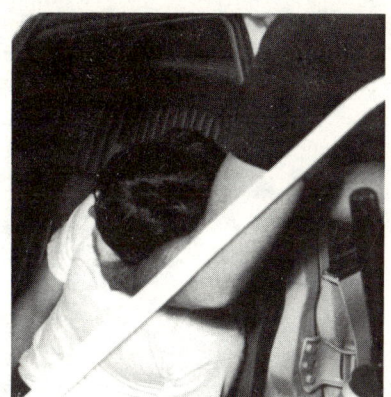

Figure 16-21

If resistance is still met at this point, the left hand can be repositioned through sliding motion so that the fingers of the left hand are still on the chin of the subject and can still supply pressure against the head and neck. However, from this position, the left thumb can be placed in the mandibular angle area of the left ear and excruciating pain can be added to the neck tension to the degree that eventually results in compliance in the form of vehicle exit. (See Fig. 16-22)

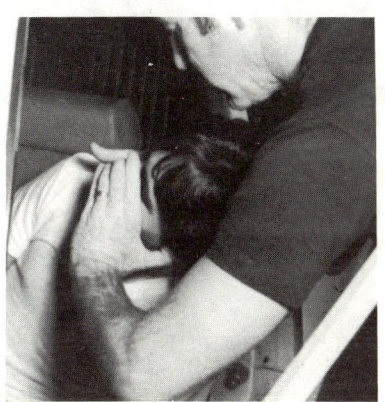

Figure 16-22

Although the head/neck twist technique has perhaps the longest history of utilization by officers on uncooperative drivers, it has been placed third due to the fact that it only removes drivers from their position and does not result in a controlled position for handcuffing. In the author's experience, the takedown/handcuffing combination most adaptable to this removal technique is Form 4. Usually the subject will raise at least one, if not both, hand to his head in order to remove the officer's grip on the chin. Once the subject is off the seat of the vehicle, the removal has all but been accomplished and submission and control should be the officer's primary concern. (See Fig. 16-23)

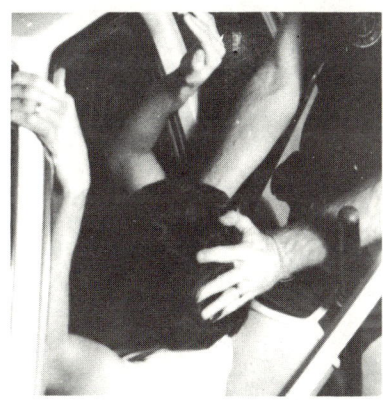

Figure 16-23

The subject will be distracted by his loss of control and the fear of falling to the ground. Therefore, direct your attention to either of the hands closest to you that would facilitate the Form 4 outward palm twist and back pressure, and control the subject's eventual exit. (See Fig. 16-24, Fig. 16-25 and Fig. 16-26)

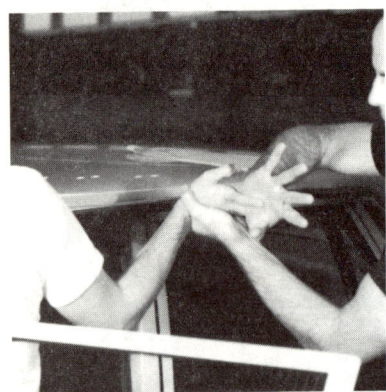
Figure 16-24

Figure 16-25

Figure 16-26

Maintaining the grip via the Form 4 application allows increased tension to secure and maintain compliance. (See Fig. 16-27)

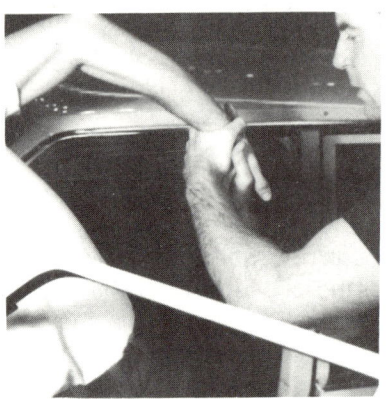
Figure 16-27

Chapter 17

WEAPONLESS COMEALONGS

On occasion, an officer may desire to subdue and control a subject and yet not desire to move into a handcuffing posture. Included in this lesson are several techniques that accomplish the implied purposes of a "comealong" hold; control, plus immediate removal capability. It must be stressed at the outset that these techniques are simplified for our learning presentation, but they require hours of practice to become viable for street utilization.

FORM 4 (REAR WRIST LOCK)/GOOSE NECK (TRADITIONAL)/GOOSE NECK (REVERSED)/ADAPTATIONS

Rear Wrist Lock

The basic Form 4 position has been established by the officer via one of the previously mentioned techniques or by effecting the following approach. In this technique, the officer has gained a position slightly to the rear of the presumed strong side of the subject. The officer initially grasps the right hand (fingers) of the subject with his right hand (See Fig. 17-1), and immediately begins to twist the lower arm inward while creating an upward elevation of the lower arm to a point of right angle with the upper arm. (See Fig. 17-2)

Figure 17-1

Figure 17-2

Once the subject's right hand has neared or achieved maximum palm exposure, the officer's left hand locks onto the blade edge of the exposed hand with the fingers wrapped around the hand edge, eventually laying across the subject's right palm. (See Fig. 17-3)

Figure 17-3

223

The officer's left thumb is positioned on the back of the subject's right wrist. The right hand can assume a handshake position to reinforce the controlling tension and gain a pattern of cooperative response. (See Fig. 17-4)

Figure 17-4

Simultaneous to the sumbission tension created on the wrist of the subject, the officer begins to direct the subject's controlled arm to the rear. (See Fig. 17-5)

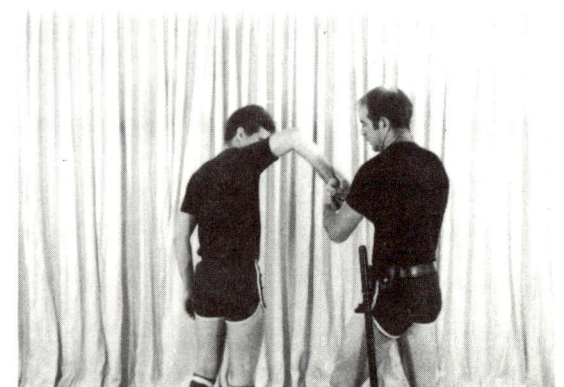

Figure 17-5

Once the bicep area of the subject's controlled arm becomes nearly parallel with the ground the officer should move rapidly to trap the elbow against his chest while simultaneously moving his right hand from the gripping position to a point where the officer's right palm applies pressure against the subject's wrist and backside of his hand. (See Fig. 17-6)

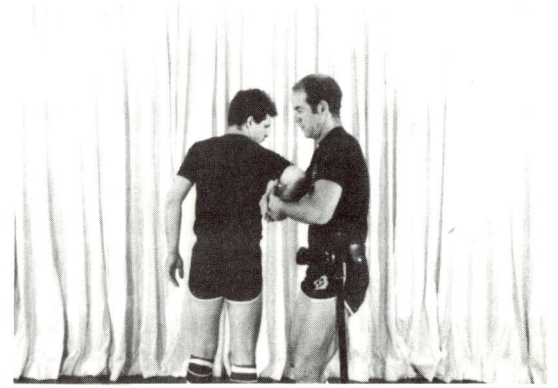

Figure 17-6

Once the final position is achieved, the subject's wrist is bent in such a manner that the fingers are pointed outward and toward the rear. The officer now moves toward the right side of the subject, trapping his controlled right arm under the officer's upper right arm and inside upper chest.

Submissive cooperation is achieved when tension is applied against the backside of the subject's wrist. (See Fig. 17-7)

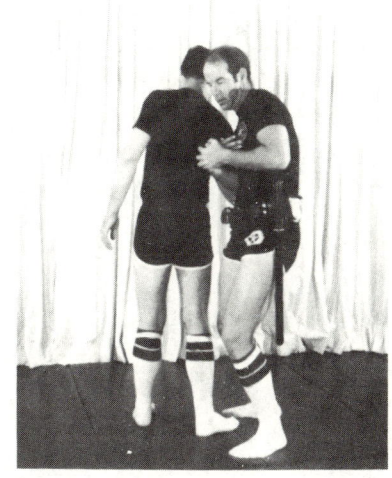

Figure 17-7

Goose Neck (Traditional)

From the rear wrist lock position, the officer has adaptive potential for at least two other operative comealongs.

The officer's left hand is placed in a palm overlay position to supplement pressure against the subject's wrist. (See Fig. 17-8) The officer now pulls the subject's wrist and forearm outward, pivoting to a parallel, forward facing position, to the rear side of the individual (See Fig. 17-9), while trapping the subject's right elbow under the officer's left upper arm and chest. (See Fig. 17-10)

Figure 17-8   Figure 17-9   Figure 17-10

In its final position, the hold situates the subject's palm toward the officer with the fingers straight down. (See Fig. 17-11) Once submission control has been assumed, the officer's strong hand could be removed from its original wrist pressure placement and freed for other uses.

Figure 17-11

Goose Neck (Rotated)

From the traditional goose neck position, the wrist posture can be changed without any control loss. The change is executed by the officer rotating the subject's controlled hand in a clockwise motion. (See Fig. 17-12)

Figure 17-12

The subject's bent wrist continues the rotation until his fingers are pointing toward his rib cage. This angle increases the tension on the wrist for even greater compliance.

Figure 17-13

This rotation technique can be continued until the subject's fingers are pointed directly upward and maximized tension is created.

Figure 17-14

Rear Armlock

While standing directly to the front of the subject or on the proximate strong side angle (See Fig. 17-15), grasp the subject's right tricep area placing the right hand between his rib cage and arm with the right hand. (See Fig. 17-16)

Figure 17-15

Figure 17-16

Simultaneous to the grasp, the officer's left forearm initially blocks the right lower arm of the subject and proceeds to force the forearm to the rear. (See Fig. 17-17)

Figure 17-17

Immediately, jerk the upper arm forward via the tricep grasp while continuing rearward pressure against the lower arm, causing the elbow to bend. (See Fig. 17-18)

Figure 17-18

At this time, rapidly pivot to the right rear/side of the individual driving your left hand under and then over the upper rear elbow area. (See Fig. 17-19) Simultaneously, capture his right wrist in the crotch of your left elbow. (See Fig. 17-20)

Figure 17-19

Figure 17-20

Now place your left forearm next to his back to prevent his wrist from escaping from the elbow crotch capture. Via this action, or the aid of an actual rear pull to the upper torso, move into a more secure rear position near the subject. (See Fig. 17-21)

Figure 17-21

Next, grasp the clothing and/or skin of the subject to gain or maintain this close control. Control tension can be established by bringing the right palm up to the lower jaw and chin of the subject, bending him backward and creating increased instability. This position also allows for nerve compression techniques including areas such as the mandibular angle, the underside of the jaw, and the base of the nose for even greater compliance. (See Fig. 17-22)

Figure 17-22

Chapter 18

ESCORTING AND TRANSPORTING PROCEDURES

Once a subject has been systematically handcuffed and searched, his potential threat to the officer has been reduced, but certainly never totally eliminated. Our recommended style of escort places the subject in an uncomfortable, unnatural, and off-balanced position, and yet allows the officer to be fully cognizant of the escort environment as well as in complete control of his prisoner.

The officer positions himself to the left rear and side of the handcuffed individual. He cautiously thrusts his left arm between the subject's left forearm and side. (See Fig. 18-1)

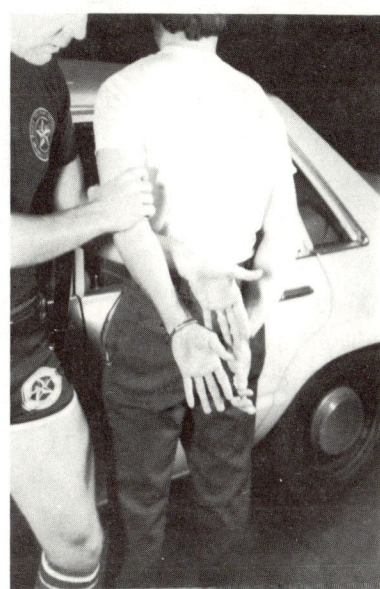

Figure 18-1

Initially, make contact with the back of the subject's left hand, gradually moving to the point where your palm applies pressure against the wrist to a state of controlling tnesion. With this tension inducement, move the subject's left elbow along your left side until it can be trapped and held with the assistance of your left upper arm. (See Fig. 18-2)

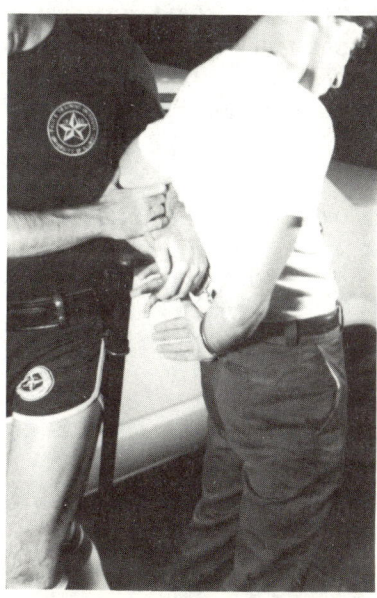

Figure 18-2

In this posture, the subject is moved backwards while the officer walks forward capitalizing on the fact that the subject remains both physically and psychologically disoriented. (See Fig. 18-3)

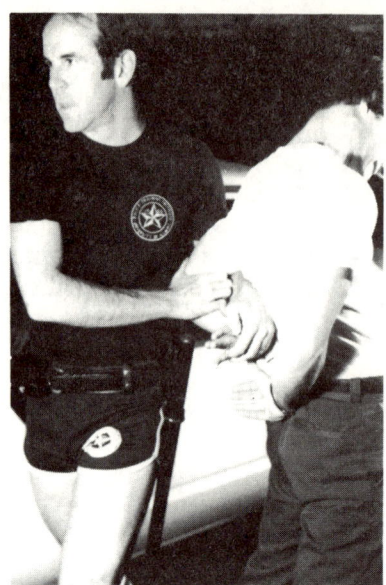

Figure 18-3

Additionally, the officer has a free hand available to open building doors or eventually the door on the transporting vehicle.

Once at the transporting squad car, walk the subject to the passenger side rear door. Open the door while you still have the wrist and arm lock on the subject. (See Fig. 18-4)

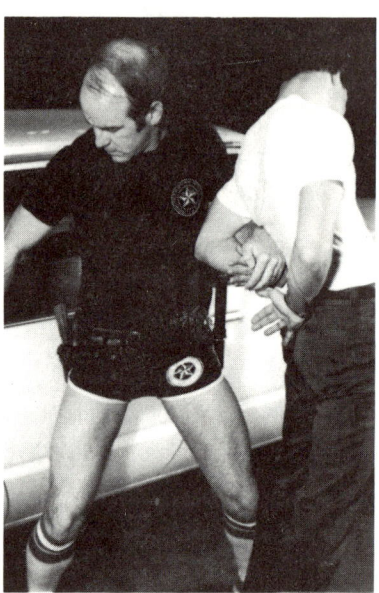

Figure 18-4

Now move your control hold from the left arm to the right arm and wrist. (See Fig. 18-5)

Figure 18-5

Remember to make the transition via the rear of the subject to prevent the vulnerability of common frontal kicks.

From this right side and front of the subject, release your right hand from the subject's wrist pressure position. Your left hand now assumes the control posture (wrist lock) on the subject's right arm. Place your right palm high on the subject's chest with your index and middle fingers extending upward and into the resion of the Jugular Notch. (See Fig. 18-6)

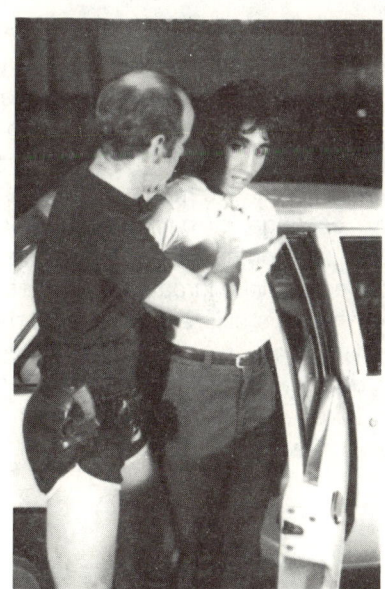

Figure 18-6

Your left palm should be placed on the upper back side of the subject's head. Advise the subject to sit down into the vehicle. The extent of fingertip pressure into the Jugular Notch area is dependent on the degree of resistance offered by the subject. Your left hand can increase the pressure to the neck region if necessary, but its primary placement purpose is to prevent the subject from hitting his head on the top or door frame of the vehicle as you push him straight back and down on the rear seat. (See Fig. 18-7)

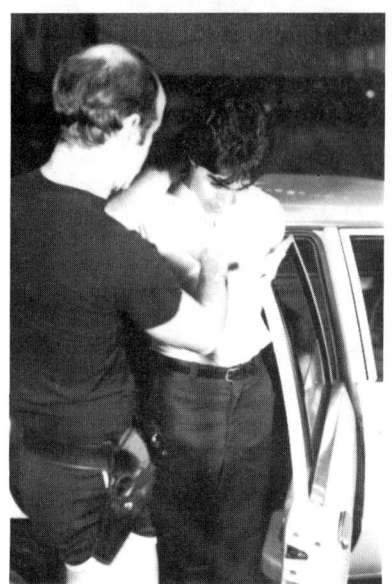

Figure 18-7

Lastly, it must be noted that the officer must keep his right leg to the outside of the subject's right leg, since a straddled position exposes the groin to a knee thrust, or other mid-torso attacks.

Once the subject has been seated, he should be advised to place his legs into the vehicle keeping his feet flat on the floor during the entire transport. Most people eventually respond to the officer's commands when they are expressed in a clear, authoritative manner; however, there must be a contingency plan for the most belligerent individual. You have positioned the subject on the rear seat with the throat pressure technique; however, his legs still remain extended out the door opening. The escorting officer cautiously closes the door to the point of its application of direct pressure against the subject's knees and shin area. (See Fig. 18-8)

Figure 18-8

During this action distracting the subject, the partner officer moves to the left rear of the squad car, opens the door and enters the vehicle.

While the subject is held in the seated position due to the leg trap, the partner officer moves to the side of the subject and initiates a compliance hold. The partner officer takes his left hand and clamps his fingers over the bridge of the subject's nose and places his thumb into the mastoid area located just under the ear against the back of the jawbone hinge. The partner's right hand wraps around the subject's head so the palm is placed on the forehead and the forearm lays along the side and rear of the subject's head. (See Fig. 18-9)

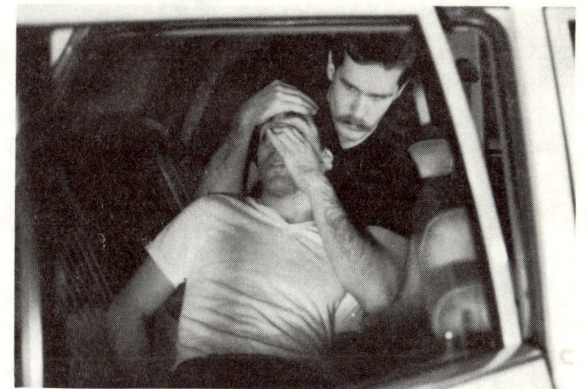

Figure 18-9

As pressure is applied to the mastoid nerve, the partner officer concurrently turns the subject's head in the direction he wants him to move or to bring about the retraction of the person's legs finally into the squad car.

TRANSPORT PROCEDURES

Seat Belt Restraint

For even greater safety to the officer and subject, and increased restraint potential, the rear seat belt should be applied to the transported individual. Pull the appropriate seat belt to its entire length outside the retractor, wrap the belt around the prisoner's wrist, then across his lap and into the securing receptacle. Remember, the seat belt can be unsnapped during the transport by the transported subject. Therefore, constant vigilance should be maintained, with frequent checks on seatbelt security, tension, and lock maintenance. (See Fig. 18-10)

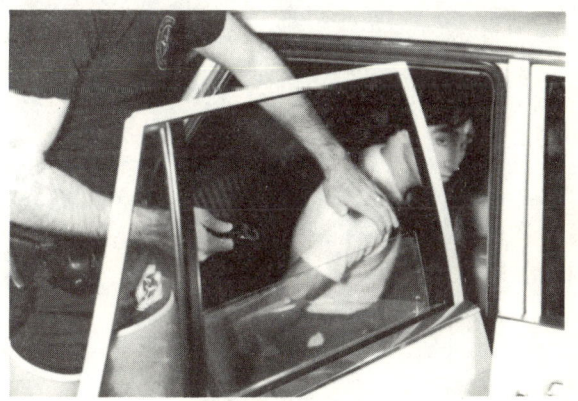

Figure 18-10

To safeguard from a head thrust attack, the officer may wish to restrain the subject's head and upper torso via a pressure application of the weak hand knife edge pressure to the nose. (See Fig. 18-11)

Figure 18-11

Officer/Prisoner Seat Placement

Since we believe in a dual officer transport system, the seating mode is simplified. The second officer should be seated to the immediate left rear position of the driver. (See Fig. 18-12)

Figure 18-12

In this position, he is best suited to counter any form of attack on the driver officer. This position is suited to safety, regardless of the number of subjects transported in the rear seat. (See Fig. 18-13)

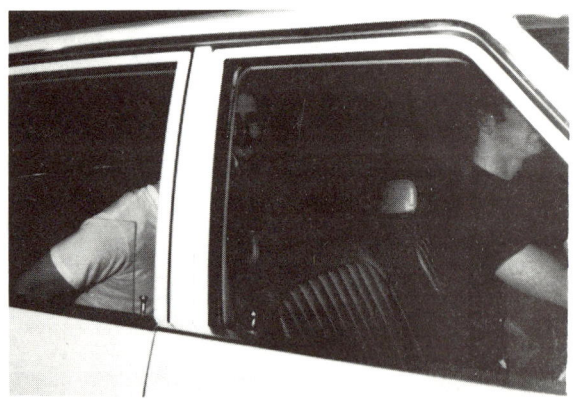

Figure 18-13

The only exception to this rule would be a properly equipped vehicle structured for either front and/or rear seat transport with a lone officer. Remember that the exception is founded upon proper equipment, not departmental or personal excuses of manpower deficits, down time, previous arrest familiarity with the prisoner, etc. Also remember that your vigilence must be heightened as a result of the 1:1 transport mode. The outside rear view mirror can be utilized for driving awareness while perhaps the interior rear-view mirror should be positioned to allow for intermittent observance of the subject. (See fig. 18-14)

Figure 18-14

Chapter 19

BATON TECHNIQUES

Without question, the baton is potentially the most versatile weapon in your personal arsenal. With proper training, it can assume the crucially needed versatility position of weapon utilization on the wide continuum of no force at one extreme and the other ultimate extreme of deadly force with your firearm.

It is our feeling that an officer should carry his baton on all calls, on all shifts. Only by its availability can it be truly effective to the officer. The baton is another responsibility you will take on as a result of our duty to consequently enhance your ability to respond.

The first lesson to be learned is that the baton not only is a tool, but is also a weapon. Even when used properly, you are likely to cause pain and possibly injury to the assailant. Circumstances will dictate when you can or should use the baton. However, unlike the firearm, the law has neglected the placement of parameters on acceptable and unacceptable applications.

Nomenclature

The baton is most often wood, with plastic and metal models growing in popularity. The time-proven wood baton is perhaps the type most widely recommended due to its cost and light weight.

The baton ranges in length from 20" to 28" with the 25" length most widely used. The terms applicable to the baton are depicted below. (See Fig. 19-1)

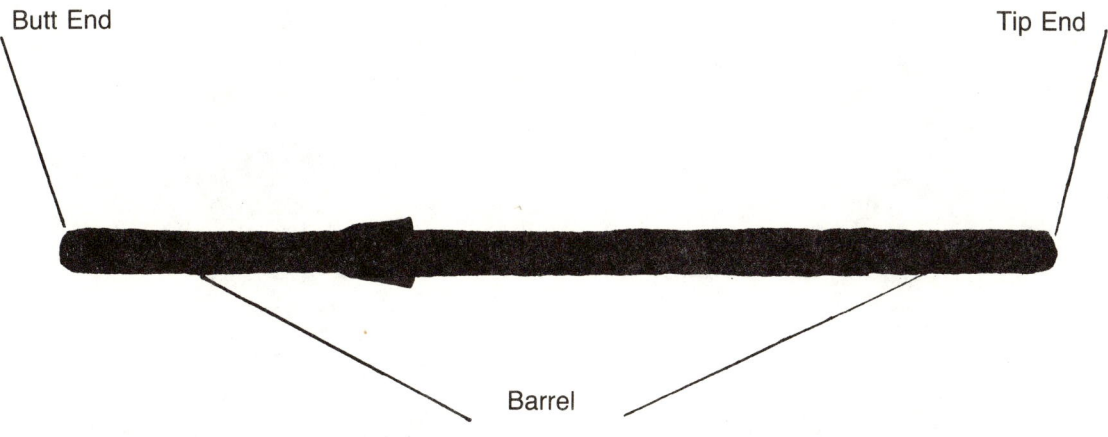

Figure 19-1

Some officers desire a leather or plastic thong to provide for a surer grip. Whatever your decision of such an inclusion on your baton, you should be aware of both the pros and cons of the thong. Rejection of the thong is generally based upon its inconsistent length due to stretching, its propensity to hook onto your handcuff case or other uniform article, door knobs, etc. Supporters of the thong profess a guarantee of a more secure grip, the capability of greater accessibility to the baton once removed from the ring holder, and a potential temporary restraining device.

If the thong is chosen, it must fit properly. The thong should be tied securely 6" to 8" from the butt end of the stick. (See Fig. 19-2)

Figure 19-2

A hole should not be drilled into the baton since it will greatly weaken the baton. To adjust the thong to fit the hand, the loop is passed over the thumb and across the back of the hand with the baton hanging below. (See Fig. 19-3 and Fig. 19-4)

Figure 19-3

Figure 19-4

The thong should be shortened until the butt-end of the baton touches the blade edge of the hand and then a firm grip established. (See Fig. 19-5)

Figure 19-5

The baton should remain in the ring holder specifically designed for the type baton selected. (See Fig. 19-6) The baton should be on the side opposite that of the officer's firearm. It should be suspended approximately 7" to 10" above the ring and the remainder below. (See Fig. 19-7)

Figure 19-6

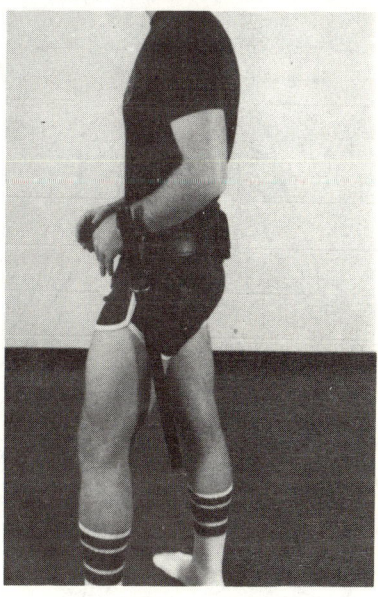

Figure 19-7

Principles of Weapon Utilization

As a general rule, the primary quality of the baton is its ability to effect a shocking blow to the assailant. Secondary capabilities of the baton are in the area of blocking kicks and punches and as a component for several comealong holds.

If the intent of your actions is to inflict a blow on the aggressor, attack those places where bone is close to the skin; i.e., elbows, wrists, knees, shins, ankles.

If circumstances allow for such action, attack priorities can and should be enacted in the following order:

1. Large muscle groups (i.e., thigh, calf, buttock)
2. Small muscle groups (i.e., forearms, biceps, triceps)
3. Exposed bones (i.e., shin, knee, wrist)

Due to increased issues of liability, officers are encouraged to target their controlled retaliation below the waist preferably at the knees, shins and ankles. This lower torso attack minimizes even accidental strikes to the head or face. *Under no circumstances should the officer target the head or face due to its high risk status.* A blow, even misdirected by the subject, could be fatal. The proper tool for the act of deadly force is your sidearm, not your baton, and definitely not your flashlight. From a purely tactical angle the head is a very elusive target and therefore may cause the officer to waste both time and technique.

In order to preserve your personal as well as professional position in enforcement, you should have a complete knowledge of the vulnerable areas of the body and avoid the targeting of these regions to prevent death or permanent injury. The drawings below clearly depict these possible fatal points of impact. (See Fig. 19-8 and Fig. 19-9)

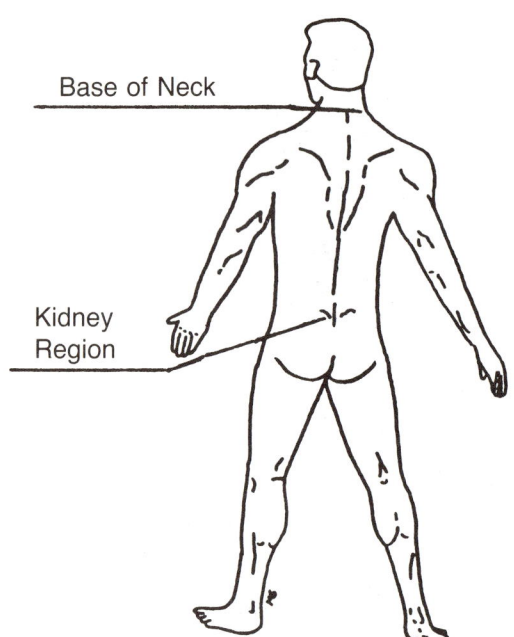

Figure 19-8

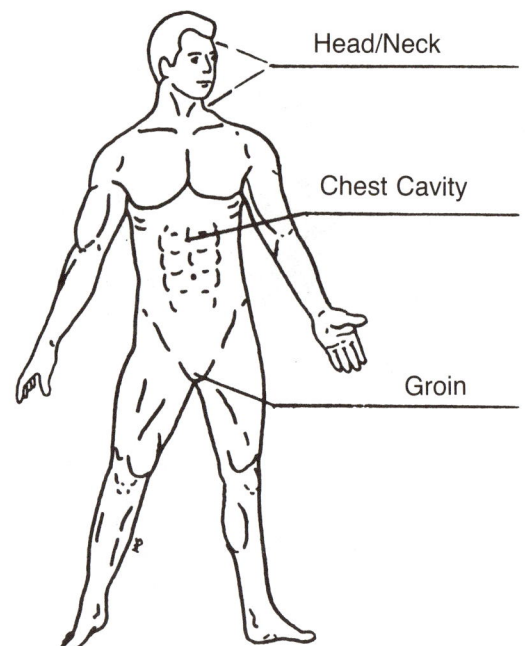

Figure 19-9

"On-Guard" Positions

In the interest of effectiveness, learning ease, and sheer simplicity, three on-guard positions will be taught. The first is a two-hand guard, the second a strong-hand guard, and the third a weak-hand guard position.

Two-Hand

The baton is held with the strong hand on the (grip end) and the weak hand placed approximately 4" from the tip end of the baton. (See Fig. 19-10)

Figure 19-10

The baton should be held relatively close to the upper chest, diagonally to the floor. (See Fig. 19-11)

Figure 19-11

From this position, the officer can protect himself from punches or kicks using the baton as a blocking or redirection source. It is readily apparent to a potential aggressor that the officer has the baton in a retaliatory position if the need develops, so as to maximize the deterrent capability of the weapon.

Strong Hand

The grip is held by the strong hand with the baton positioned to the rear of the strong leg while assuming the alert stance. The actual gripping of the baton should be loose, with control occurring as much by palm pressure against the baton and buttocks rather than a tight grasp. (See Fig. 19-12)

Figure 19-12

With this "hidden" position, the officer can escape from the possibility of precipitating violence due to an overreaction on the part of persons encountered, while in no apprciable way jeopardizing the officer's rate of response.

Weak Hand

The weak hand is loosely gripped around the baton as it remains in the ring holder. The loose grip allows for faster and more diversified direction if necessitated during the encounter. This on-guard position is recommended highly due to the fact that it appears the least antagonistic to the citizenry and yet provides a ready striking status if the need arises. (See Fig. 19-13 and Fig. 19-14)

Figure 19-13

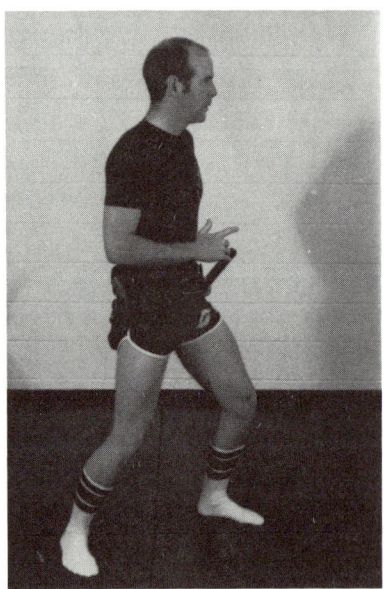

Figure 19-14

BATON BLOCKING TECHNIQUES

High Block

From the "military stance" position with the weak foot forward, step forward with the weak foot while simultaneously raising the nightstick with both hands to a point 0 above the head and in front of the face. (See Fig. 19-15)

Figure 19-15

It is vitally important that the nightstick remain at a 30-45° angle when raised, for this will deflect any blow over the hand and to one side. (See Fig. 19-16)

Figure 19-16

This also enables the officer to redirect the force of the blow to one side, instead of accepting it full force (nightstick parallel to the ground).

While stepping back to the original stance, retract the nightstick to its original position. (See Fig. 19-17)

Figure 19-17

The block is extremely quick and the officer should refrain from leaving the baton in the block position for any length of time because of the exposed nature of his rib cage.

Low Block

From the "military stance" position with the weak foot forward, advance one step with the weak foot while simultaneously extending the arms and nightstick toward the ground at a 30-45° angle. In this block, the converse is true in regards to the angle of the nightstick. The stick should be parallel with the ground and the arms at a 30-45° angle. (See Fig. 19-18 and Fig. 19-19)

Figure 19-18

Figure 19-19

A key aspect is that the officer's feet remain beneath him and that he not lean forward to accept the intrusion. (See Fig. 19-20)

Figure 19-20

If the nightstick is not low enough to suit the officer, he should broaden his step rather than bend his torso toward the ground. While stepping back to the original foot position, retract the nightstick to its original position in a smooth controlled manner.

Outside Block

From the same "on guard" position, step forward with the *opposite* leg (strong foot) and face the weak side while simultaneously extending the baton. (See Fig. 19-21 and Fig. 19-22)

Figure 19-21

Figure 19-22

In this position, the baton is perpendicular to the ground while the elbows are in the center of the body. (See Fig. 19-23)

Figure 19-23

The strong hand is closest to the ground and the feet are no more than 2 times shoulder width. (See Fig. 19-24, Fig. 19-25 and Fig. 19-26)

Figure 19-24

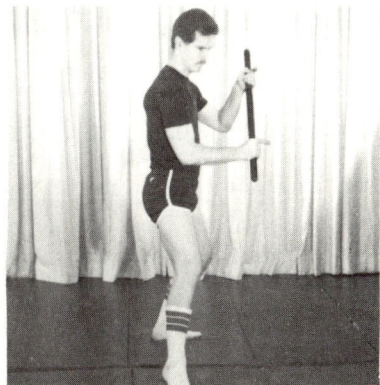

Figure 19-25

Figure 19-26

Once the block has been utilized, move rapidly back into the original ready position or initiate a sternum strike with the butt-end of the baton if necessary.

Inside Block

From the same "on guard" position, step forward with the weak foot and face the strong side while simultaneously extending teh baton. (See Fig. 19-27 and Fig. 19-28)

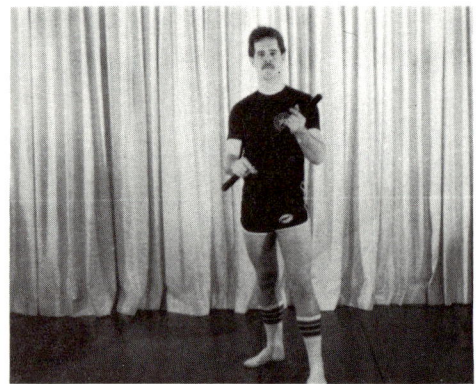

Figure 19-27

Figure 19-28

Again, the baton is perpendicular to the ground while the elbows are in the center of the body. The strong hand is closest to the ground and the feet are no more than 2 times shoulder width. (See Fig. 19-29 and Fig. 19-30)

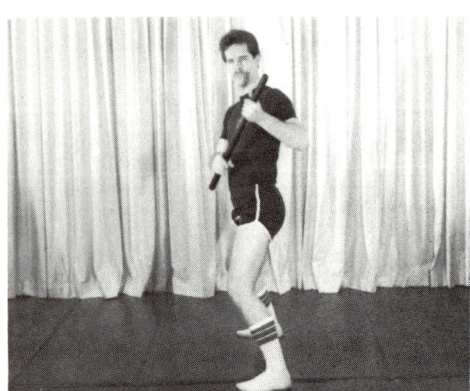

Figure 19-29

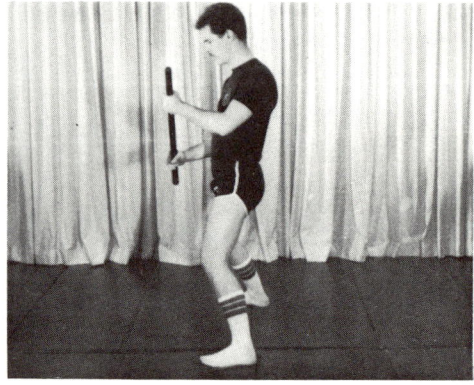

Figure 19-30

Once the block has been utilized, move rapidly back into the original ready position.

Fan Block

With the nightstick in the strong hand behind the strong leg and with the weak foot forward, step forward with the weak foot at a 45° angle from the original position and bend the strong leg. (See Fig. 19-31)

Figure 19-31

Simultaneously, rotate or fan the baton in an upward movement, that is parallel to the body, in a semicircle. (See Fig. 19-32, Fig. 19-33, Fig. 19-34 and Fig. 19-35)

Figure 19-32

Figure 19-33

Figure 19-34

Figure 19-35

249

Return to the original position upon completion of the block. (See Fig. 19-36)

Figure 19-36

Blocks in Action

To enhance your understanding of prescribed blocking techniques the following photographs provide a pictorial representation of blocking techniques in a simulated confrontation.

HIGH BLOCK

Figure 19-37

Figure 19-38

LOW BLOCK

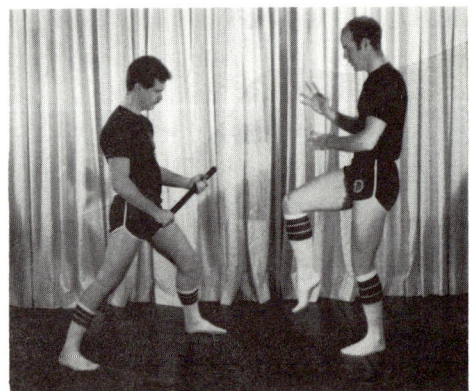

Figure 19-39

Figure 19-40

250

OUTSIDE BLOCK

Figure 19-41

Figure 19-42

INSIDE BLOCK

Figure 19-43

Figure 19-44

FAN BLOCK

Figure 19-45

Figure 19-46

Three-Count exercise

Upon the commencement of the assailant's attack, initiate the first count from the weak-hand on-guard position. Move your weak foot forward in a sliding motion drawing the baton butt first from your ring holder targeted toward the sternum of the subject. (See Fig. 19-47 and Fig. 19-48)

Fig. 19-47

Figure 19-48

Remember to position the weak hand properly on the baton to secure the strike. The three fingers should wrap around the outside of the baton, the index finger should lay flat on the top of the baton to secure and sight the thrust, and the thumb should secure the interior portion of the baton and minimize any horizontal movement of the baton on impact. The baton must be held close to the forearm at the point of strike to prevent injury to the officer during the sequence. (See Fig. 19-49)

Figure 19-49

At the time of impact, capitalize on the reactive momentum of the baton by moving your strong hand to a gripping position approximately 2" from the tip of the baton, with the palm of the strong hand outward. (See Fig. 19-50)

Figure 19-50

The baton should move in a circular motion down past the belt line and continue around back up to the shoulder level of the officer. (See Fig. 19-51 and Fig. 19-52)

Figure 19-51

Figure 19-52

The baton, still held in the two-hand grip should proceed horizontally across the officer's upper chest directly into the subject's sternum. (See Fig. 19-53 and Fig. 19-54)

Figure 19-53

Figure 19-54

Again, continue the reverse motion of the baton on impact, backstroking the baton across the officer's upper chest continuing the reverse circular motion down past the belt area, upward to the officer's lower chest region. (See Fig. 19-55 and Fig. 19-56)

Figure 19-55

Figure 19-56

Once the baton reaches this position, it should initiate a two-hand thrust to the previous target area of the sternum. (See Fig. 19-57 and Fig. 19-58)

Figure 19-57

Figure 19-58

Some important factors should be highlighted to maximize the effectiveness of the techniques. First, stay flat-footed, sliding the feet to assure full contact with the ground and a constant balanced state. Make the actual thrusts in a snappy movement to minimize defenses and prevent the baton from being grabbed by the assailant. For similar reasons, never lean out to strike and jeopardize your balance status. Make each total movement of the baton in a circular continuation with the baton held as close as possible to the body to prevent someone on the side of the officer from grabbing the stick. (See Fig. 19-59 and Fig. 19-60)

Figure 19-59

Figure 19-60

Defensive Strike Exercises

In this technique, the officer assumes a stance that is only broken due to the aggressive persistence of the assailant. The officer remains in the normal alert stance in reference to the lower torso. The baton is previously removed from the ring holder and gripped in the strong hand. The baton is placed in a tucked position behind the weak upper arm and shoulder. The majority of the barrel of the stick should be hidden due to the folded, overlapping position of the arms. To the observer the officer should appear totally relaxed with approximately 3" of the butt of the stick below the weak elbow while 4" or more of the tip and barrel extend above the weak shoulder. (See Fig. 19-61)

Figure 19-61

Once the attacker places the officer in a vulnerable position, three things should happen simultaneously: (1) the weak arm should be raised with the forearm positioned above the officer's head to protect from an overhand strike and to distract the assailant, (2) the strong foot should be slid back to lower the entire body to strengthen the officer's fighting stance and cast the appearance of a truly defensive situation if observed by any witness; and (3) the "cocked" baton should explode from the tucked position downward targeted at the knee region of the subject's lead leg. (See Fig. 19-62, Fig. 19-63, Fig. 19-64 and Fig. 19-65)

Figure 19-62

Figure 19-63

Figure 19-64   Figure 19-65

Remember, this technique is passive and not designed to move subjects out or away from their unlawful position. This technique is utilized to allow the officer to "hold the line" if necessary or simply to resume a strong defensive position where sufficient time allows for the assumption of the position.

Chapter 20

SPECIALIZED BATON TECHNIQUES

BATON ISOLATION TECHNIQUES

Particular street situations may call upon the officer to remove a particular individual away from the scene in a safe and expedicious manner. For instance, a particularly abrasive person is beginning to create a disturbance in a bar, an angry relative attempts to "resolve" a family dispute involving his sister, or perhaps an angry female protester refuses to leave the entrance to city hall. Of course, all of these situations and participants could be allowed to run full course and mandate an arrest. However, let's say that if the violence prone catalysis, the patron at the bar, the self-righteous relative, or the female activist were removed, the entire situation would defuse itself. Such activity and benefits suggest proper dispersal techniques.

Each of these techniques allow the officer to remove a subject to a nearby location with increased control via the baton.

Groin Lift

Gain a position to the rear of the subject. With your weak hand placed mid-distance on the baton, carefully thrust the tip and barrel of the baton between the legs of the subject. (See Fig. 20-1 and Fig. 20-2)

Figure 20-1

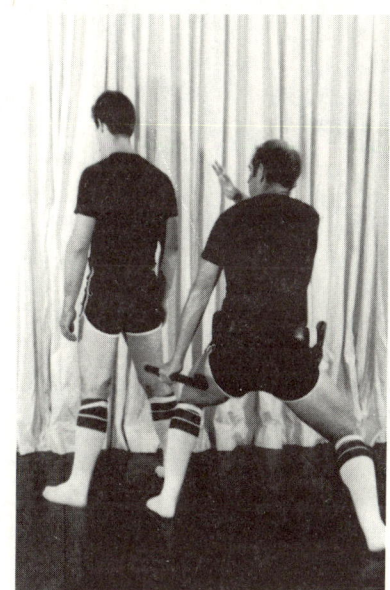

Figure 20-2

257

The baton should be controlled so as to cause a short debilitating reaction by the subject by sheer shock contact with the baton, but with no actual injury potential. Once the controlling hand has passed beyond the leg arch, twist the stick so that it is now parallel with the floor and providing a pressure block to the subject's upper thigh region. (See Fig. 20-3 and Fig. 20-4)

Figure 20-3

Figure 20-4

Simultaneously, gain control of the subject's upper shoulder or neck region via either clothing, hair, or skin tissue. (See Fig. 20-5)

Figure 20-5

Move the person forward by lifting the baton and subsequently initiating a forward projection of the subject who is now experiencing a loss of balance and limited lower torso control. (See Fig. 20-6)

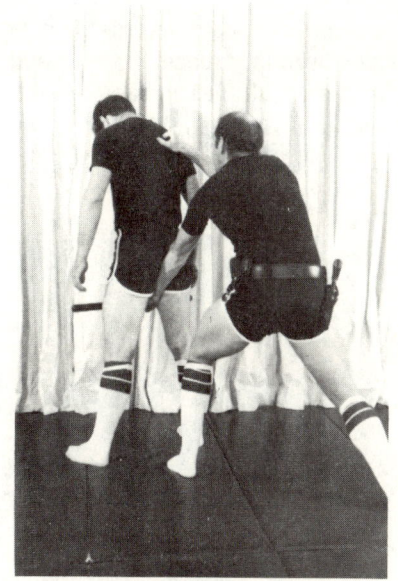

Figure 20-6

Remember that you must maintain constant control over the forward motion of the subject through your lift and push. If for any reason the subject becomes uncontrollable, pull out of the groin lift posture and regain an on-guard position. (See Fig. 20-7 and Fig. 20-8)

Figure 20-7

Figure 20-8

Although variance does exist, it is felt that when the lifting arm is in position, the hand should be postured so that the knuckles are down and the palm is up. (See Fig. 20-9)

Figure 20-9

This hand posture minimizes the injury risk to the officer should the subject suddenly sit down. Also, if the subject were to overpower the officer and pull the baton from his hand, the officer can still inflict a devastating attack upon the groin area. (See Fig. 20-10)

Figure 20-10

Those who support the palm down technique indicate that it insures a superior grip on the baton, virtually negating the possibility of the subject forcing the baton from the officer's grip. (See Fig. 20-11)

Figure 20-11

Additionally, this grip style allows the officer greater lifting strength since the upper arm and shoulder muscles are working in combination. Proponents of the palm down lastly state that if control is lost in the employment of the technique, this grip style facilitates quick dispersal or removal of the weapon so that the possibility of the officer's wrist being trapped is mitigated. (See Fig. 20-12)

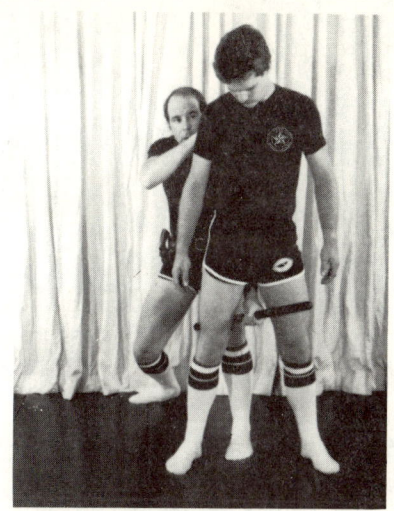

Figure 20-12

Based upon practice and experience, the technique choice should be individually selected.

Armpit Shove

Gain a position where you can grasp the subject's left hand in a handshake position, or a combination of the index, middle or ring fingers. (See Fig. 20-13)

Figure 20-13

Immediately, invert the palm, raising the arm upward, exposing the left armpit region as the left arm is raised. (See Fig. 20-14)

Figure 20-14

261

Lay the tip end of the baton into the left armpit, keeping constant forward pressure upon this highly sensitive area. (See Fig. 20-15)

Figure 20-15

While the left wrist is still bent back and the baton tip is strategically placed, advance deliberately, moving the subject the desired direction and distance. (See Fig. 20-16)

Figure 20-16

BATON DOMINATION TECHNIQUES

As those circumstances of encounter increase in officer directed hostilities, the baton can act as a source of immediate, effective control.

Radial Lock

Gain a position to the right side of the subject with the baton in a stature of right hand readiness. With a distracting blow to the right rib cage or interior of the right elbow, the right arm should move to the side and rear. (See Fig. 20-17)

Figure 20-17

Simultaneous to this movement, thrust the tip and barrel of your baton through the arch of the armpit. (See Fig. 20-18)

Figure 20-18

Pivot to the right rear of the subject raising your own left arm up and over the tip of the baton, catching and securing it between your upper left arm and side. (See Fig. 20-19 and Fig. 20-20)

Figure 20-19

Figure 20-20

Move your left arm up between the subject's right arm and side, trapping his right wrist so that it is resting on your forearm. (See Fig. 20-21)

Figure 20-21

Control the subject's right elbow by keeping it snug to your rib cage. (See Fig. 20-22)

Figure 20-22

To effect the degree of tension inducive of submission, lift up with your elbow while pushing down with your left hand. (See Fig. 20-23 and Fig. 20-24)

Figure 20-23

Figure 20-24

This situates the baton against the radial bone and the associative nerve impulses. It should be added that tension can be best enhanced if the subject's right hand is positioned so the palm is pointed downward maximizing the direct contact of the radial bone and baton.

Shoulder Torque

Gain a position to the left front angle of the subject while the baton is held in the right hand. Initiate a distracting strike to either the suspect's left rib cage or interior left elbow area. (See Fig. 20-25)

Figure 20-25

Advance with the left foot while simultaneously thrusting the tip and barrel of the baton between the subject's left arm and rib cage. (See Fig. 20-26)

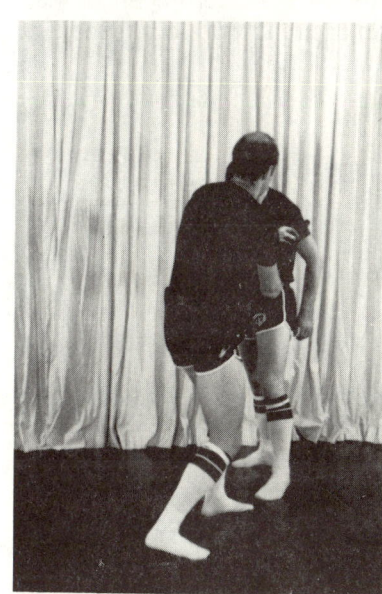

Figure 20-26

Grasp the tip of the baton by taking your left hand and reaching over the subject's left upper shoulder area. (See Fig. 20-27)

Figure 20-27

Once this grip of the baton tip is established, pivot to the rear side of the subject. From this position of leverage, the suspect's left arm is now trapped at a 90° angle in reference to the upper arm and forearm and the baton enhances the torque action into an arm and wrist lock. (See Fig. 20-28)

Figure 20-28

The subject can have the tension increased, and therefore an enhanced state of submission by grasping clothing in the right rear shoulder of the suspect with the officer's left hand. In this position, the suspect can now be pulled toward the front of the officer, increasing the torque on the forearm and wrist of the baton controlled left arm via a wrist lock technique. (See Fig. 20-29)

Figure 20-29

Chapter 21

WEAPON RETENTION TECHNIQUES

The dilemma of weapon retention is one more disturbing degree of the criminal environment you are facing. In fact, now you see the necessity of a dual posture of protection; your personal defense as well as your service revolver.

Each year the viciousness of the criminal increases against the officer as evidenced by these statistics from the FBI Law Enforcement Officers Killed summary. Nationally, from 1975 nearly 20% of the officers killed, were killed with their own service revolvers.

Our techniques are founded upon three preventative maxims requiring constant diligence on the officers part:

1) Primary and periferral vision must be maximized to detect any hostile threat. The distance between you and the suspect and your constant attitude of alertness are the best "guarantees" of safety.

2) Draw your weapon primarily for the obvious purpose . . . firing!! Once the weapon is drawn and not utilized, for whatever reason (improper assessment of the threat, lack of legal justification, etc.), you have dramatically increased the criminals chances for disarming.

3) The most secure place for the weapon prior to its utilization is the holster. If the holster is properly designed for the weapon; is free of "trick draw" vulnerabilities, and is capable of stabilization via our weapon retention techniques, it remains the safest refuge for the service revolver.

For learning simplicity, we have divided our retention techniques into three phases. These phases must, and will achieve effective fluidity, once practiced sufficiently by the officer.

Phase I: Secure the Weapon

At the onset of the intervention into the officer's zone of protection, he must instinctively react to secure the weapon in the holster. At this point, the issue of attitude must also be "set"; "set" on the goal of survival. You must seize the initiative for the subsequent action.

Phase II: Assume a Position of Advantage

Now that the weapon is secure, you must move into a position that allows for maximized diffusion against the attacker, continued, if not an enhanced, protection of the service revolver, and to set up the third phase of weapon release and subject debilitation.

Phase III: Weapon Release and Subject Debilitation

Procedurally, the weapon has been made secure, the officer has moved into a posture of effective counter action, and now an application of physical stress is directed toward the subject, checking the attempted weapon removal and effecting a temporary debilitation.

PRIMARY GRIPS FOR RETENTION

Strong Hand Grip

Bring your strong hand up behind your holstered weapon. (See Fig. 21-1)

Figure 21-1

Your strong index finger wraps around the top of the hammer, your palm covers the entire trigger guard area, while the entire hand grasps the weapon into a holster fixed position. (See Fig. 21-2)

Figure 21-2

It is felt that this primary method of securing the weapon can be most applicable to the wide variety of holsters marketed. If, for instance, the holster utilized is not a "break-away" style, downward pressure can additionally be provided, which will add to the retention effort.

Weak Hand Grip

Bring your weak hand across the front of your body. (See Fig. 21-3)

Figure 21-3

Your weak thumb wraps across the top of the hammer, your weak index finger wedges in behind the trigger, the rest of your palm and fingers cover the trigger guard area, while your entire hand grasps the weapon into a holster fixed position. (See Fig. 21-4)

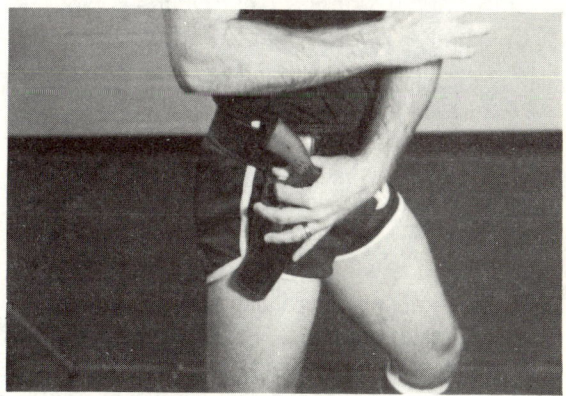

Figure 21-4

Frontal Retention Technique

The following is an example of a frontal retention technique. Although a specific example is pictorally provided below, the officer should keep in mind that this technique is merely one of a myriad of techniques that may be used depending on the officer's preferences and the situations in which they are utilized. To reiterate, the officer should keep in mind that any type of technique that is used should contain the three phases of the retention system previously indicated.

Upon the subject's initial attempt to disarm the officer, the first movement is to secure the weapon in its holster as described in Phase I. The officer steps back with the strong leg to assume a defensive posture that not only sets up a debilitation technique, but provides security for the service revolver. (See Fig. 21-5 and Fig. 21-6)

Figure 21-5

Figure 21-6

Notice that upon the officer stepping away from the subject, the service revolver is behind the officer at the furthest point from the subject. This fading away technique places the officer between the service revolver and the subject as advised in Phase II.

The final procedural aspect of this technique is the debilitation effort. The technique applies a forearm strike to the back of the subject's elbow to achieve the weapon release. (See Fig. 21-7)

Figure 21-7

Note, although the technique is separated into each of its discrete parts for instruction, it should be done in a continuous motion. In other words, once the weapon is secured in the holster (Phase I), the fading away and the elbow strike may be simultaneously applied (Phase II & III). The more fluid the technique is applied, the more effective it will be.

Rear Retention Technique

Upon the subject's initiation of a disarming attempt, the officer should immediately act to secure the weapon in its holster as indicated previously. (See Fig. 21-8)

Figure 21-8

Once the weapon is secured, the officer should begin to move to a defensive posture or a position of advantage.

Upon gaining a position of advantage, the weapon release may be achieved by a variety of debilitation techniques. Depicted below, the officer may utilize an elbow strike or hammer fist strike to the wrist or the elbow. (See Fig. 21-9 and Fig. 21-10) Notice that each of the phases of the retention system is observable and followed.

Figure 21-9

Figure 21-10

271

SPECIALIZED RETENTION TECHNIQUES

Elbow Compression Technique

Upon the initiation of a disarming technique, the officer assumes a modified weapon securing posture. (See Fig. 2-11) This is done by compressing the elbow and tricep area against the weapon to pin it against the holster, and consequently, the hip area. Notice that Phases I and II are combined in this technique. The officer is fading away and securing the weapon simultaneously. (See Fig. 21-12)

Figure 21-11

Figure 21-12

To achieve the weapon release, again, a variety of debilitation techniques may be employed. As depicted below, the officer selected a palm strike to the chin and backfist to the face. (See Fig. 21-13 and Fig. 21-14)

Figure 21-13

Figure 21-14

Both or either of the techniques may be employed by the officer depending on the circumstances or the severity of the assault.

Rear Elbow Compression Technique

Upon the attempt to disarm the officer, the elbow compression technique is an appropriate instinctive reaction of the officer. This technique may be employed for both front and rear disarming attempts. Notice the lowering of the center of gravity by the officer while employing Phase I. (See Fig. 21-15 and Fig. 21-16)

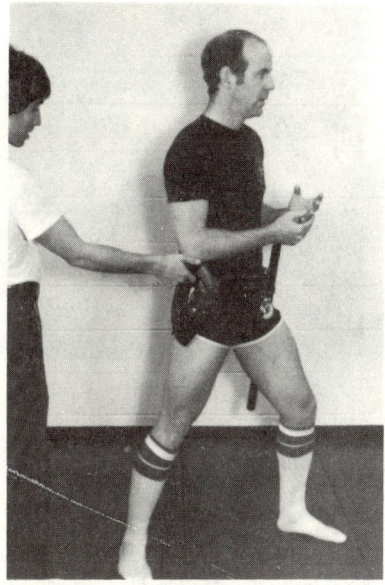
Figure 21-15

Figure 21-16

Upon the achievement of the security of the weapon, the officer rotates 180 degrees to achieve a defensive posture and initiates a series of debilitating blows including a palm heel strike to the face and a palm instep kick to the shin. (See Fig. 21-17 and Fig. 21-18)

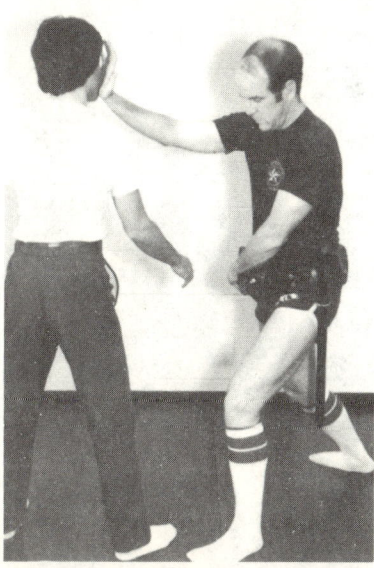
Figure 21-17

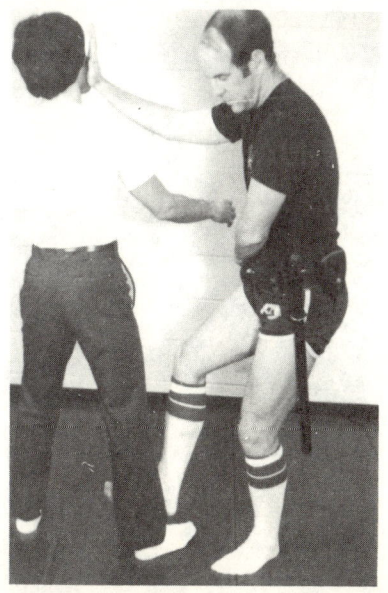
Figure 21-18

Assuming this defensive posture also achieves the weapon release. In many instances, by merely assuming a defensive posture, the weapon release will be achieved.

Down Retention Technique

To maintain weapon security, the officer should lie on the side of the weapon (weapon next to the ground) to place himself between the subject and the weapon. (See Fig. 21-19)

Figure 21-19

The officer should attempt to raise himself to an elbow supported lying position to maintain a clear view of what is occuring.

From this position, the officer should employ kicking techniques to ward off the subject's attack including but not limited to the knee, groin, or lower back area. (See Fig. 21-20, 21-21, and Fig. 21-22)

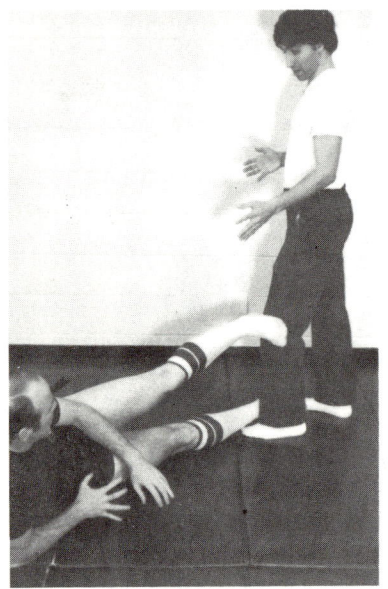  

Figure 21-20   Figure 21-21   Figure 21-22

274

The officer should remain in this position until assistance arrives or he gains an advantage. This advantage may be achieved by a well placed kicking strike to the subject which temporarily stuns him and allows for insured success in standing. The officer should remember that he is better off remaining in the down fighting position if he cannot successfully achieve a standing position.

Chapter 22

DOWN FIGHTING TECHNIQUES

A police officer will undoubtedly find himself in personal defense confrontations in which the subject is dominating. The situation may appear as though there is no type of advantage to be gained to control it and defensive success is futile. In such situations, traditional methods of personal defense are inappropriate. Consequently, not only is it important to learn perspectives and techniques for the obvious types of confrontations officers will encounter, but every aspect of personal defense confrontations must be dealt with in order to offer a comprehensive personal defense program. Since there exists feasible situations in which the officer will be in an extreme disadvantage, these must be pursued to offer exposure, as well as a course of action, to assist in the personal defense of the officer in such confrontations. These situations may manifest themselves in circumstances such as losing one's service revolver, being knocked to the ground by an unanticipated blow, or ending up on one's back during the course of a confrontation. Each of these types of situations are extremely feasible occurances during the course of a police officer's career. Regardless of whether or not a situation appears hopeless, the officer should use his defensive training and knowledge to attempt to salvage the situation in some fashion. Whether this will involve merely cooperating with the subject in order to find an advantage previously undiscovered, or attempting to apply a defensive technique immediately, the officer should keep in mind that it is HIS perrogative only. The salvaging of the situation is based upon the *officer's* assessment of the circumstances and his faith in his ability to successfully gain an advantage or defend himself. There are, however, a few concepts and techniques that may assist the officer in these "futile" situations. These concepts and techniques dwell in the realm of merely surviving, as opposed to controlling, the situation.

"Down fighting" refers to a situation in which the officer finds himself on his back attempting to defend himself. Although the usual response of most police officers is that of denying even the remotest possibility of such an occurance, situations such as this can, and do, happen. The next response of most police officers is that if they find themselves in such a position, the first course of action is to get to their feet. This too, is equally as erroneous as denying the possibility of the situation. There exists a science and an art of defending oneself while in a prone position—the science of down fighting. Consequently, the proper perspectives, as well as techniques, warrant the attention of the officer, for the mastering of these skills may preserve the officer's safety.

Down fighting, to reiterate, is considered a last ditch effort to defend oneself without the use of deadly force. It incorporates all the knowledge and skills previously learned in personal defense. For instructional simplicity, the perspectives and skills have been divided into phases that should be strictly adhered to by the officer.

Phase I: Pre-Prone Perspectives and Techniques

In the pre-prone phase of down fighting, the officer is completely aware of the physical confrontation. The confontation may have progressed to a point where a variety of techniques may have been attempted but have proven ineffective. During the course of defending himself, the officer may have either lost his balance on his own or with the assistance of the subject. Consequently, the first aspect of the pre-prone phase has manifested itself—a loss of balance by the officer. This loss of balance is of critical importance for without the balance loss, no down fighting techniques should be attempted. The officer should utilize traditional personal defense techniques unless he is FORCED to use down fighting skills. Simply put, the officer should not resort to down fighting techniques in lieu of traditional personal defense unless he has no option.

The next critical aspect of the pre-prone phase is in regards to the balance loss. If the balance can be maintained WITHOUT jeapardizing total personal defense, the officer should do everything possible to remain on his feet. A crude example of this concept is that if an officer has lost his balance and is falling forward while the subject is attacking from the rear, it would not behoove the officer to try to catch himself if that entails sustaining injury from the rear assault. To attempt to catch one's balance in this situation would involve increasing the risk of sustaining injury from the attack. Not only would the officer reduce the amount of time before the attack (because he would be moving TOWARD the attack), but he would increase the force of the attack (physics of collision). It is imperative that the officer assess the situation accurately and respond in such a fashion that promotes his safety. In the previously described scenario, the best course of action for the officer to take would be to go ahead and fall forward and increase the amount of time before the attack. This increase in time would allow the officer to attempt to regain a defensive posture prior to the attack. It would also diffuse the force of the attack because he would be moving away from the attack instead of into it.

The officer should keep in mind that regardless of his decision of whether to attempt to maintain balance or to fall, once the decision has been made, it should be followed. One of the most intense repercussions in personal defense is a concept referred to as cognitive dissonance. Cognitive dissonance is indecision. Cognitive dissonance results in no action. If the officer does not know whether to fall or attempt to maintain his balance, he will, in effect, do nothing. Doing nothing promotes injury. Cognitive dissonance manifests itself most clearly in situations in which, for example, the officer decides to go with the fall and at the last second, attempts to pull out of it. Greater risk of injury is found when cognitive dissonance occurs than when it doesn't. If the officer has COMMITTED himself to take a particular course of action, it should be followed through. Simply put, if the officer decides either to go with a fall or to attempt to maintain his balance, he will be in less jeapardy than if he initiates one action and changes his mind. Additionally, an incorrect decision—if followed through—is less hazardous than no decision.

Another important concept in the pre-prone phase of down fighting is falling. If the officer cannot maintain his balance, the primary issue in his mind should be that of the fall. The fall, by itself, may eliminate the possibility of forced utilization of down fighting techniques. The falling techniques that were previously practiced by the officers are specifically applicable to the pre-prone phase. If the officer falls forward, he should attempt a forward roll out of the fall and immediately assume a defensive posture. This will eliminate the down fighting situation as is desired. The same holds true for falling backwards. The officer should roll out of the fall and again, assume a defensive posture. The down fighting position should be avoided if possible and this is the most viable technique to use if balance cannot be maintained.

Phase II: The Prone Phase

The prone phase is the most critical phase of down fighting. It requires the officer to follow pre-established priorities in order to maintain his safety. The first priority of the officer is to lay on his strong side. (See Fig. 22-1)

Figure 22-1

Laying on the strong side has a variety of assets. Primarily, if the officer stays on the strong side, he is protecting the service revolver. (See Fig. 22-2)

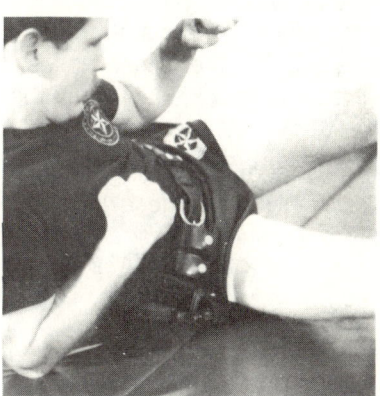

Figure 22-2

Although laying on top of the service revolver is not the most comfortable position from which to defend oneself, it is the safest. The service revolver should be as inaccessable to the subject as possible. This position insures weapon security.

The next aspect in the prone phase are defensive techniques. The primary techniques utilized in down fighting are kicks. Kicks will ward off the subject's attack and keep the adversary an appropriate distance away. While in the prone position, the officer should attempt to raise himself to his strong elbow to allow for increased visibility. (See Fig. 22-3)

Figure 22-3

One leg should be held approximately knee high to the subject, while the other leg is shin high. (See Fig. 22-4)

Figure 22-4

Note that the strong leg does NOT lay on the ground. Both legs must be used to defend oneself. Since the officer is on his strong side, the groin should not be susceptable to attack because he is on his SIDE. If the officer makes the mistake of lying on his back, the groin will be extremely susceptable to attack. Insure that the SIDE is on the ground and not the back.

The most effective kicking techniques for personal defense in this situation are the side thrust and the heel thrust. These kicks should be directed at the subject's knees, groin, and shin in singular or combination modes. (See Fig. 22-5 and Fig. 22-6)

Figure 22-5

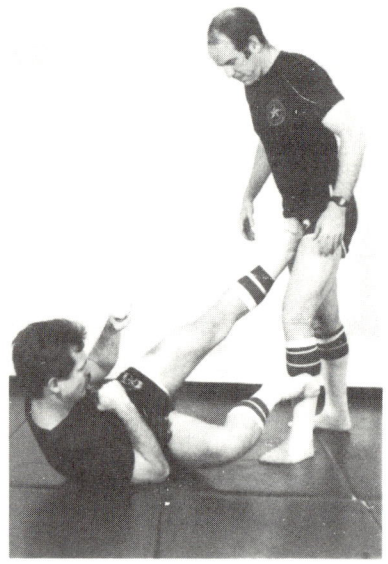

Figure 22-6

If the subject bends forward at the waist attempting to lower himself to assist in his attack, a kick may be directed at the face. (See Fig. 22-7 and Fig. 22-8)

Figure 22-7

Figure 22-8

Finally, the officer will find that if, while kicking, he raises his hip off the ground, increased power and extension will occur. The officer's hands may be used for assisting balance while kicking.

Another significant aspect of down fighting is the rotation. Rotation of the officer's body should occur if the subject attempts to move around to the officer's head. The feet of the officer should be the closest object to the subject. At no time should this position be compromised. If the subject attempts to move around to the head of the officer, he should pivot on the cheek of the buttocks to keep the subject in front of his feet. The rotation should be in a direction such that the officer is between his service revolver and the subject. Again, the officer should keep his revolver as far away from the subject as possible. (See Fig. 22-9, Fig. 22-10 and Fig. 22-11)

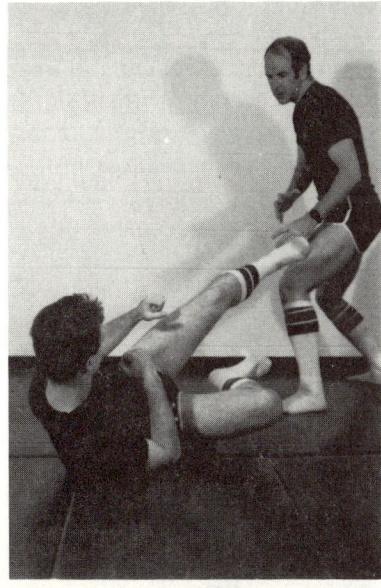

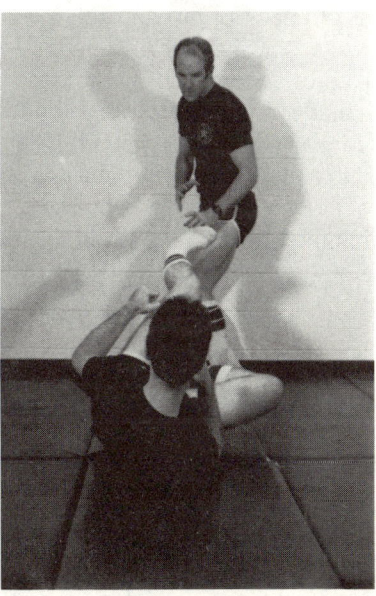

Figure 22-9

Figure 22-10

Figure 22-11

Phase III: Regaining Situational Equilibrium

Following the establishment of peronal defense and weapon defense, the next priority is that the officer regain some semblance of "control" of the situation. The issue at this point, however, is that the officer should accomplish this endeavor without jeopardizing already established defensive "control". It does not behoove an officer to keep an attacker away by down fighting techniques if he or she relinquishes this control while attempting to stand up. Although the next critical priority surfaces at this point, the message is that defensive protection intrinsically implies a certain degree of control. One may contend that while in the down fighting position, the officer is not in control. This is erroneous, for a prerequisite of protection is control. Onecannot exist without the other. The officer must have (at least) control of his own physiological reactions to successfully defend himself or herself. Consequently, the issue now, however, is the type of control being addressed.

While the officer is in control of his personal defense, he is not necessarily in control of the situation. The officer "owns" the control of his physiological responses, he does not "own" the control of the responses of others. "Controlling the situation" is a concept that is irrelevent to personal defense, in a strict sense. The personal defense of officers in confrontation situations is based upon an underlying scheme involving actions and reactions, moves and countermoves. Even if the subject has been restrained, there is no point in a confrontation in which the officer has exclusive control over the actions of the subject. The actions warranted appropriate by the officer are based upon a variety of guidelines (legal and moral) as well as situational variables. Consequently, since the focus of control of the actions of the officer is dependent upon predominantly external facets or factors, there exists no point at which the officer legitimately owns the control of the confrontation. An important point now surfaces. While the officer cannot own the control of the confrontation physically, he can, however, own it mentally. This ownership involves the establishment of alternatives if one proves ineffective, the sequencing of techniques, pre-planned courses of action, and the numerous amount of defensive perspectives previously described.

This exemplifies the conceptual framework of law enforcement personal defense. Each of the concepts are interdependent and global in nature. Each concept may be applied to every conceivable situation or confrontation involving personal defense. The conceptual approach to personal defense should not be viewed as discrete perspectives limited in scope and application. Universal application of defensive concepts is imperative for defensive success. In an obvious sense, limiting the application of concepts and techniques, limits one's ability to successfully defend himself or herself.

Upon the withdrawal of the initial attack of the subject, the officer should immediately assume a position that improves visual range or visibility and defensive capabilities. This procedure will incorporate a variety of sequences to accomplish this endeavor. Each discrete sequence of this task is designed to maximize defensiveness as well as enhance technique accomplishment. This may be accomplished if the officer remains in the down fighting position and raises himself to the strong elbow. This position will improve the analytic abilities of the officer by improving visibility. Following this modification, if the attacker has ceased the attack, the officer should make attempts to regain a standing position. This may be accomplished by following the outlined steps.

The officer now moves from his protective down fighting stance bringing his interior foot close to his buttocks. (See Fig. 22-12, Fig. 22-13 and Fig. 22-14)

Figure 22-12   Figure 22-13   Figure 22-14

At this point he rolls his hips upward, lifting the buttocks directly over the interior heel and foot establishing eventually a strong low kneeling stance. (See Fig. 22-15 and Fig. 22-16)

Figure 22-15   Figure 22-16

From this kneeling posture the officer slowly rises to his feet never jeopardizing the control, strength, and protection of his position throughout the equilibrium regaining effort. (See Fig. 22-17 and Fig. 22-18)

Figure 22-17

Figure 22-18

Chapter 23

NERVE COMPRESSION CONTROL TECHNIQUES

In order to better utilize the product of nerve compression controls one must have at least a basic understanding of its process. Of all the systems within the human body the nervous system is the most highly developed and indeed most complex.

Perhaps it is easiest to think of the nervous system as the body's communication network. The brain and spinal cord are dispatching sources while the nerve fibers are the message transmitters throughout the body.

From this profile one can and should begin to recognize what consists as the body's foundation of control. The subject who has been placed under arrest has used this communication network with its ability to act upon external as well as internal stimuli to decide if he will succumb to the arrest . . . or resist. Obviously the most critical factor for this decision is the nature of the stimuli and its interpretation. The officer's action or inaction is the most important variable that precipitates the subject's course of action. Did the officer gain control by his tactical procedure; his authoritarian presence, his spatial positioning, etc., or did the subject violently resist the arrest due to the officer's arrogant attitude, his excessive force utilization, or his image of vulnerability, etc.

Regardless of the reaction and its basis, the brain is the first and foremost element of control. One must realize in some situations the officer's actions could be entirely correct from a control perspective but the arrested subject has decided to initiate resistance. Now, however, if this occurs the officer has an alternative via nerve compression techniques to alter that decision by "tapping in" on the communication network.

Once again let's focus on the process or mechanisms involved in the system. The basic unit of the nervous system is a nerve cell or neuron. It consists of a cell body with extending threadlike processes called dendrites and axons. The dendrites are numerous, bush like in formation, and generally surround the cell body. The axon is usually long and thin and singular in number.

The axon of each cell body or neuron forms a single nerve fiber. Certain fibers in turn connect various neurons to each other. The axon of one cell extends toward the dendrites of another cell. The junction between the axon of one nerve cell and the dendrites of another is called a synapse. (See Fig. 23-1)

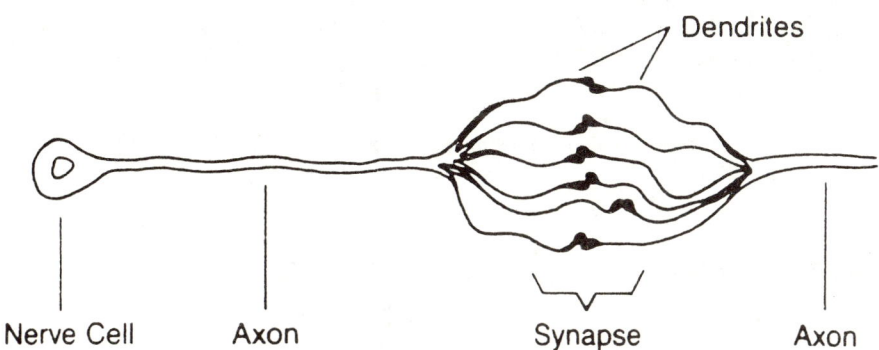

Figure 23-1

Bundles of these nerve fibers or axons coming from various nerve cell bodies, lead from the brain and spinal cord. They make up what we eventually refer to as nerves, and extend the communication system to all parts of the body. Ordinarily the brain initiates an impulse through the nerve fibers to a muscle and a response occurs.

But more specific to our concern is when a nerve is stimulated by a disturbance (pressure) along its conductor membrane (See Fig. 2) In fact, nerve compression points are nerve locations which because of their configuration and/or external accessibility are subject to direct disturbance stimulation. Nerve compression control is based upon a pressure induced electrical imbalance which either interrupt the original brain messagew to the muscle or overrides that message with a pain compliance stimulus to the brain. For instance, the subject may initially direct his body to resist an officer's control attempts, however once the officer utilizes a nerve compression control tactic the mind re-assesses its primary action and now responds to another more immediate stimulus designed to negate this primary resistance and therefore subject control is established.

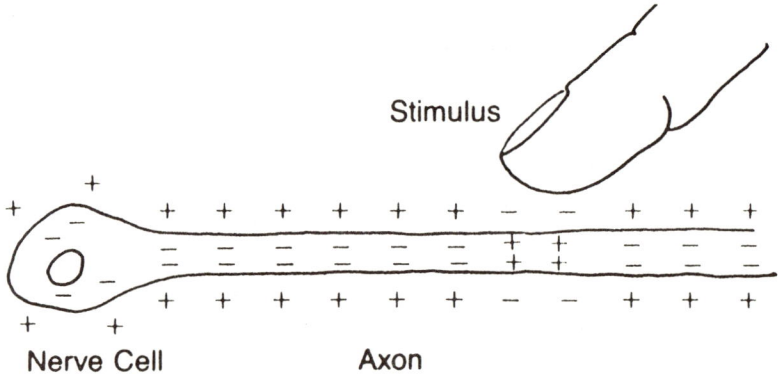

Figure 23-2

The scientific explanation of nerve energy and its human kinesiology had rather humble beginnings. For centuries rather brutal research was conducted upon human subjects. Slaves or captured prisoners were experimented upon and their potential for precise body pain profiles were documented. These illustrations of vulnerability and tactical superiority have become control resources in and of themselves.

As with most other police tactics, nerve compression controls have made their transition into modern law enforcement. These areas of the body are recognized as uniform sites, which when placed under pressure result in predictable motor reactions, generally equally responsive and susceptible to pain, and require a minimum amount of applied pressure.

As a result, nerve compression control tactics can be utilized in at least four tactical areas. One avenue of control is by using the nerve compression location to temporarily distract the aggressor. Distraction accomplishes three goals:

a. It diverts the subject's actions from offensive to defensive.

b. The initial aggressive motor action will be debilitated.

c. The subject becomes vulnerable for the application of other control methods.

Another control avenue is that of pain compliance as a result of nerve compression utilization when enhanced with specific, loud, verbal commands designed to overload the aggressor's brain until compliance is achieved.

Balance displacement can be achieved with nerve compression to restrain, restrict, or readjust an aggressor's mobility. Once we have the subject's balance we have facilitated greater potential for control.

Additionally, each of these nerve compression controls can be effected by two board categories of application: contact/penetration pressure or striking/impacting pressure. Contact/penetration includes controlled progression force as well as unarmed, quick penetration into any such area of susceptability. Striking/impacting pressure encompasses weaponless strikes to body areas and also weapon assisted impact focusing.

Before identifying specific nerve compression control areas and corresponding tactics it is necessary to understand several factors relating to pain compliance. As noted, pain presents the avenue of control and eventual compliance. Pain compliance must fall within controlled perameters so that the pain does not become brutal or sadistic. If such a condition occurs, the subject may initiate an uncontrollable level of resistance and a subsequent survival reaction.

These techniques are certainly not the panacea controls free of effective variance.

There are some physiological factors that influence the degree of pain compliance achieved, including:

1. The threshold levels of pain (pain tolerance) unique to the individual.

2. The general health and fitness level of the subject.

3. The body composition of the subject, including bone structure, nerve exposure, etc.

4. The influence of chemicals, alcohol, drugs, etc. which can act to block normal patterns of pain transmission to the brain, triggering a normal body reaction.

5. The precise area of application; wrist, fingers, ankle, each of which present variance in a person's reaction.

So too, the individual may trigger emotional responses relating to the psychological influences of pain compliance, including:

1. What was the mental state of the subject? Was he passive, mildly indifferent, or violently aggressive?

2. The ability of the individual to practice mind control techniques. This phenomenon allows the person to, in effect, disassociate a normal body reaction to pain.

3. The nature of the subject who innately exhibits minimal reaction to pain, not a practiced mental regime, but as a natural mechanism.

And lastly, but perhaps most importantly, the essence of the majority application and the moderate guarantees of effectiveness relate to the officer's utilization including:

1. Correct placement and commencement of technique.

2. The initial training as well as the duration and regimen of practice by the officer.

3. The physical and emotional level of the officer initiating the control application assures maximum control via nerve compression design and focus.

Nerve compression techniques can be maximized in effectiveness following these basic application steps:

1. *Secure grip* When utilizing direct compression tactics the targeted areas must be secured with a firm grasp.

2. *Stabilization* Once the affected area is under a firm grasp the rest of the subject's body must be stabilized for full endurance of the technique as well as a more predictable re-action once application is initiated.

3. *Precision compression* When compression is directed, the more precise the source, i.e. finger tip rather than pad, hand edge rather than palm, etc., the more intense and isolated the stimuli. Compression stimulation should be maximized via "absorbed" rather than abrupt initiation of impact.

4. *Auditory assist* Simultaneous to the initiation of a nerve compression point control, loud, concise vocal commands assist in the overriding of the brain's resistance directive. The vocal stimulus should be dual in design. First, auditory alarm can be created by an explosive shout or command. Once this has commenced, loud, concise directive messages should follow enhancing efforts to master and maintain control of the subject.

NERVE COMPRESSION FOCUS AREAS

The following nerve compression regions on the body have proven effective in enforcement activity:

Mandibular Angle Nerve Area

Three nerves, the hypoglossal, vagus, and glossopharyngeal travel together behind the mandible at the base of the earlobe. (See Fig. 23-3)

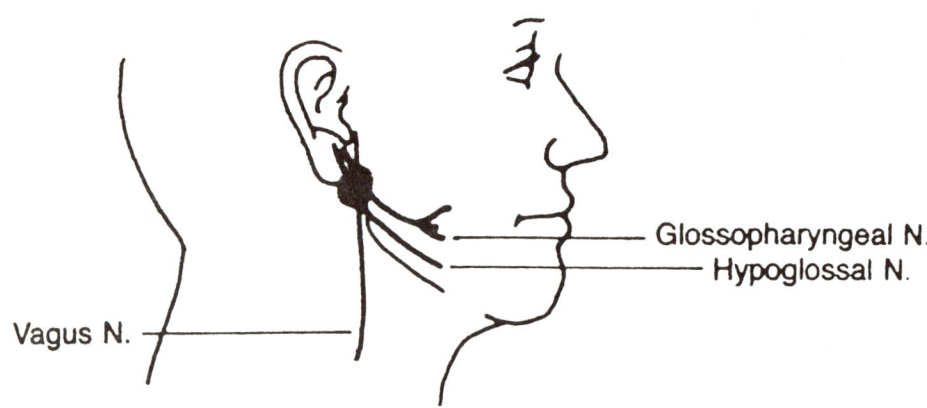

Figure 23-3

Compliance is achieved by direct compression applied behind the mastoid and mandible at the base of the earlobe inward toward the center of the head and then on an angled plane toward the nose. (See Fig. 23-4 and Fig. 23-5)

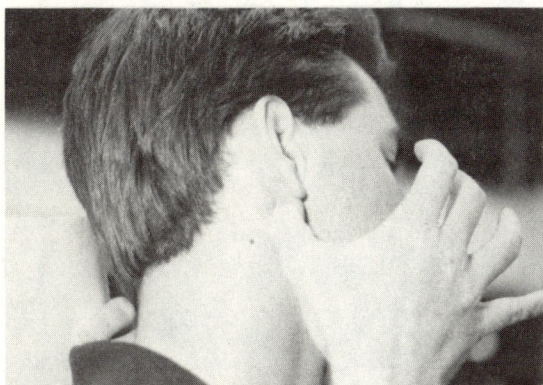
Figure 23-4

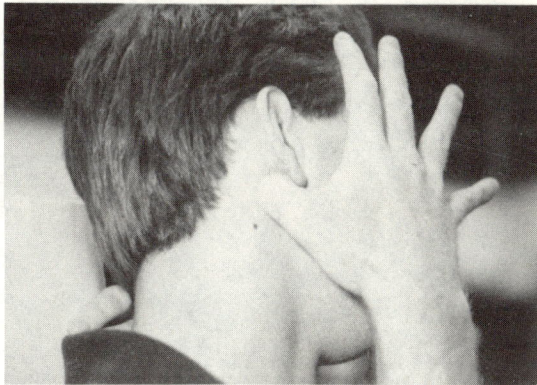
Figure 23-5

Hypoglossal Nerve Area

The most sensitive and broadly accessible nerve in the head and neck is the hypoglossal. The nerve has its origin near the ear and goes through the jaw region to the posterior of the tongue. (See Fig. 23-6)

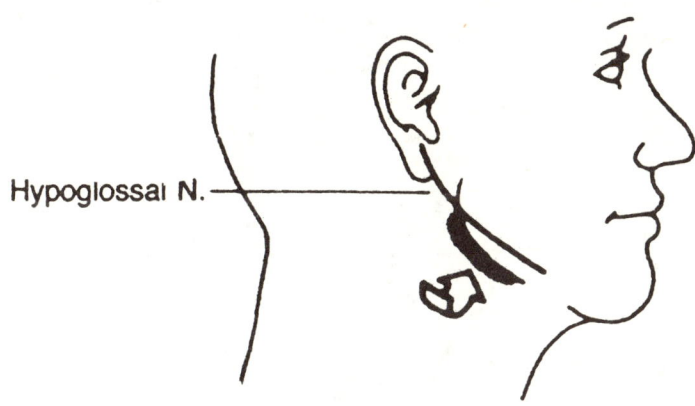
Figure 23-6

Compliance is initiated by direct compression under the jaw upward toward the center of the skull. (See Fig. 23-7 and Fig. 23-8)

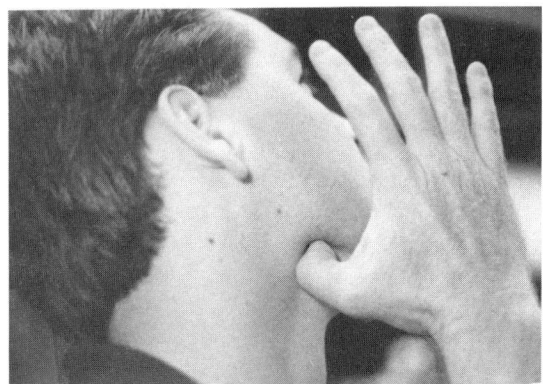
Figure 23-7

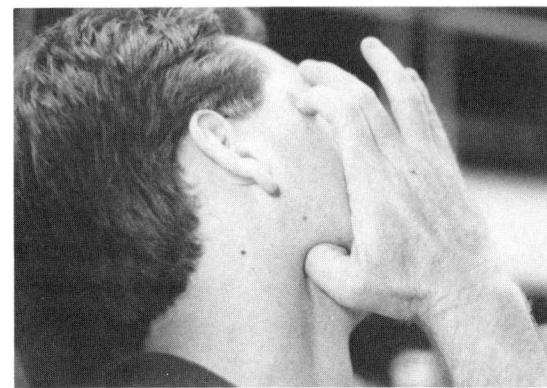
Figure 23-8

Infra-Orbital Nerve Area

This nerve area consists of the infra-orbital nerve which runs directly across the base of the nerve. (See Fig. 23-9)

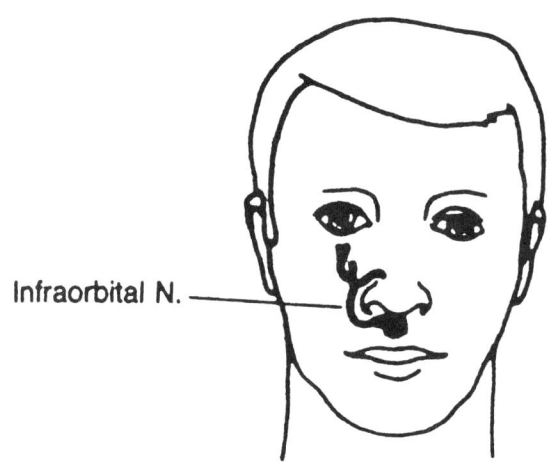

Figure 23-9

Compliance is achieved by compression application at nose base toward the upper/center of the skull. (See Fig. 23-10)

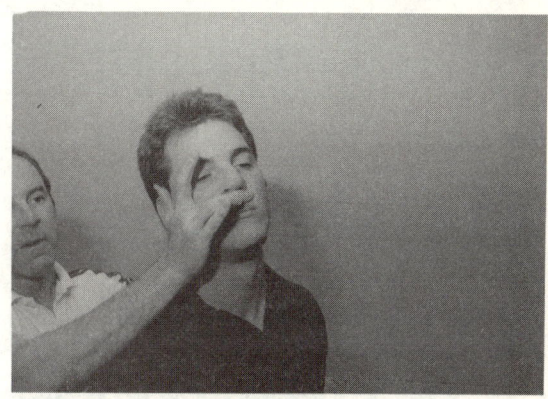

Figure 23-10

Jugular Notch Area

In this body location, two nerves, the superior laryngeal and recurrent laryngeal are affected when compression is initiated. (See Fig. 23-11)

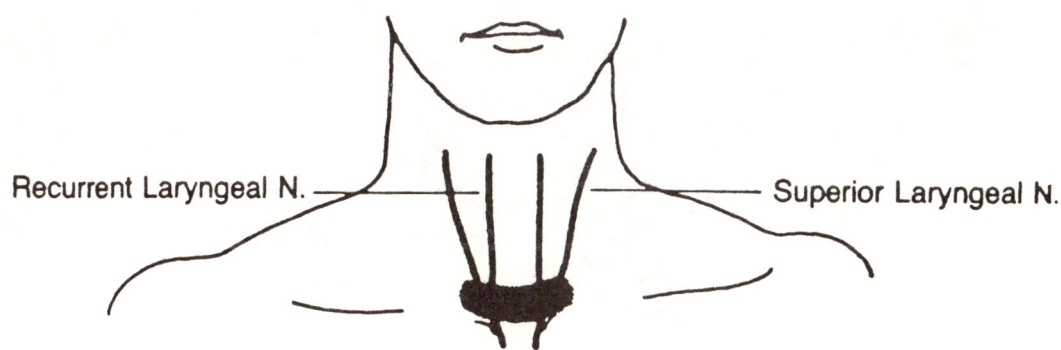

Figure 23-11

Balance displacement as well as compliance is achieved when compression is directed inward toward the center of the body on a 45 degree angle. (See Fig. 23-12 and Fig. 23-13)

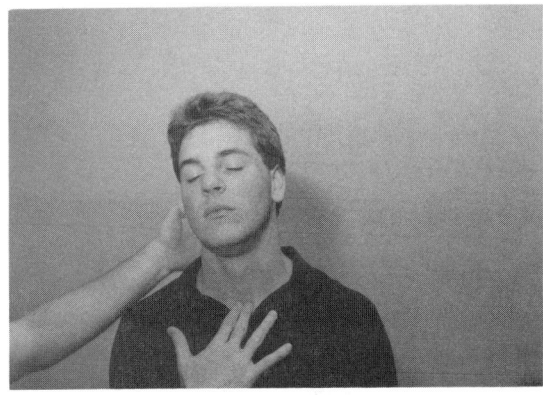

Figure 23-12

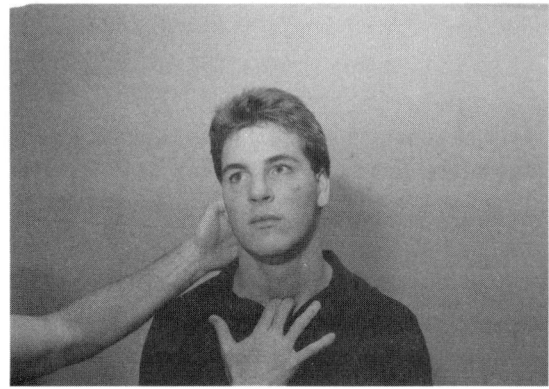

Figure 23-13

Brachial Plexus Nerve Area (Origin—Mid Neck Region)

The brachial plexus group, the median, radial, and ulner nerves, form and flow along the side of the neck approximately by 4-6 inches from the base. The origin area encompasses the mid-neck region. (See Fig. 23-14)

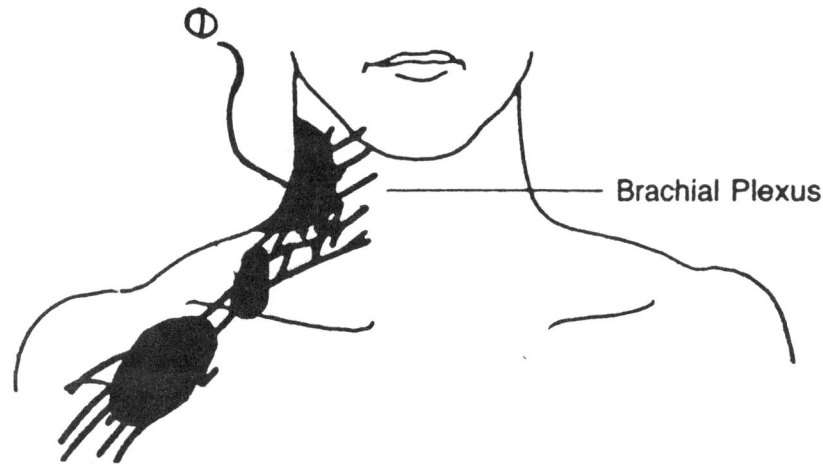

Figure 23-14

Direct compression should be directed toward the center of the neck until compliance is realized. (See Fig. 23-15 and Fig. 23-16)

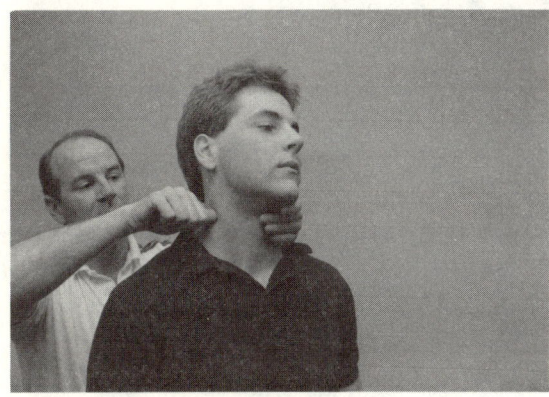
Figure 23-15

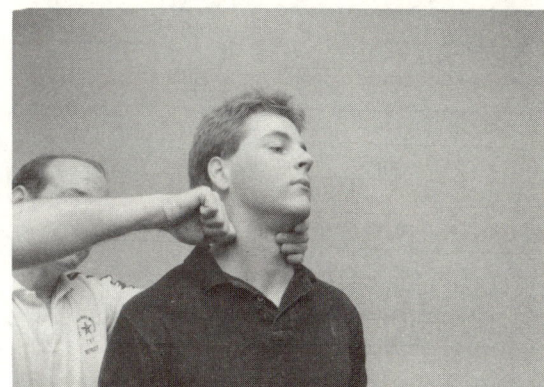
Figure 23-16

Brachial Plexus Nerve Area #2 (Clavicle Notch Region)

The clavical notch region of the brachial plexius group is the focus of area applications. (See Fig. 23-17)

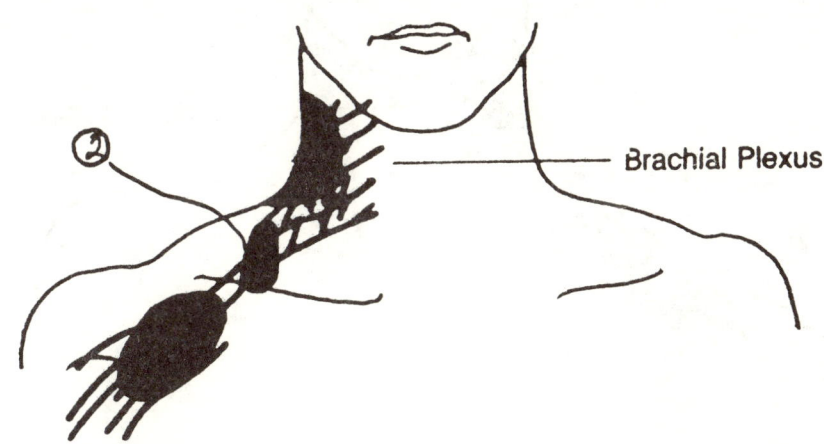
Figure 23-17

293

Direct compression should be applied downward via a 45 degree angle toward body center. (See Fig. 23-18 and Fig. 23-19)

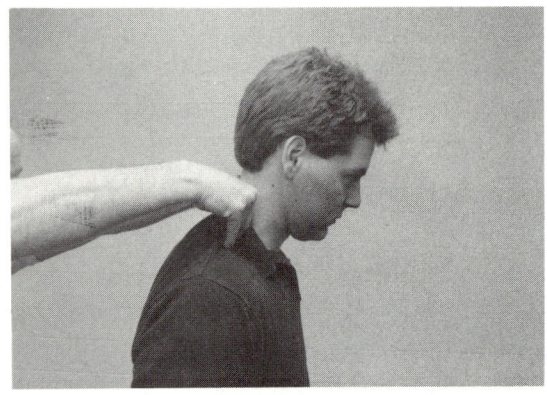

Figure 23-18

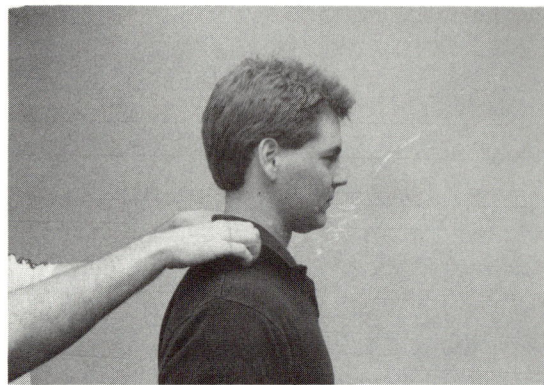

Figure 23-19

Brachial Plexus Nerve Area #3 (Deltoid/Pectorales Major Region)

The deltoid/pectorales major region is the focus for area three application. (See Fig. 23-20)

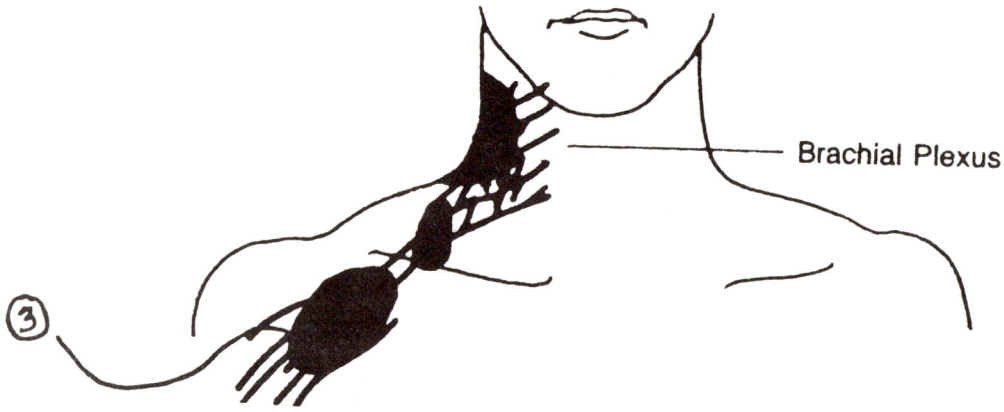

Figure 23-20

294

Once the subject's arm is extended outward, direct impact compression is effective when applied via striking techniques focused impact compression primarily toward the deltoid/pectorales major configuration above the crease of the arm. (See Fig. 23-21 and Fig. 23-22)

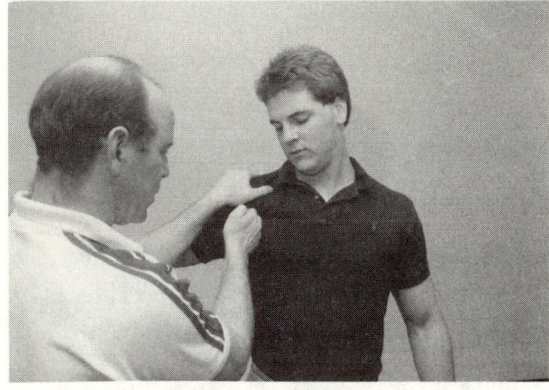
Figure 23-21

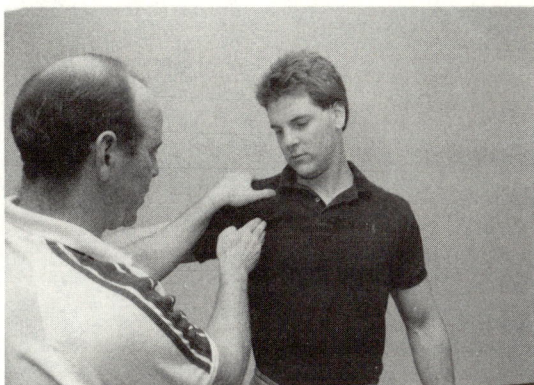
Figure 23-22

Radial Nerve Area (Top of Forearm)

One of the more exposed areas for the radial nerve is approximately two inches below the elbow break on the top of the forearm. (See Fig. 23-23)

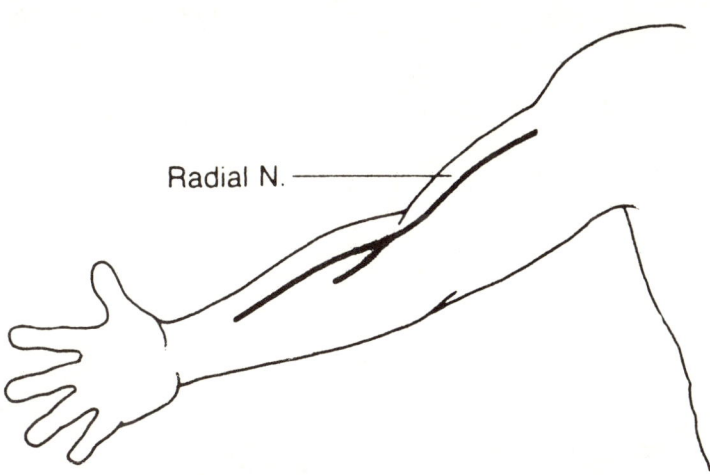

Figure 23-23

295

Compliance is achieved by contact compression or impact penetration on the target site. Impace focus should be directed to the center of the forearm in either application. (See Fig. 23-24 and Fig. 23-25)

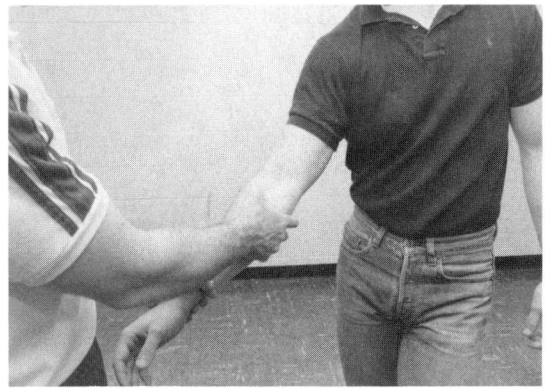

Figure 23-24

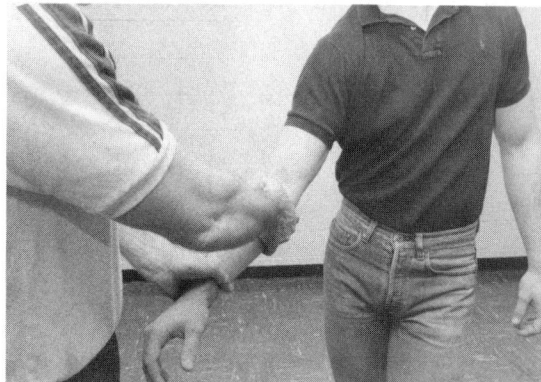

Figure 23-25

Common Peroneal Nerve Area (Outside Knee Region)

The common peroneal nerve branches off the sciatic nerve and flows along the outside of the knee region just under the kneecap. (See Fig. 23-26)

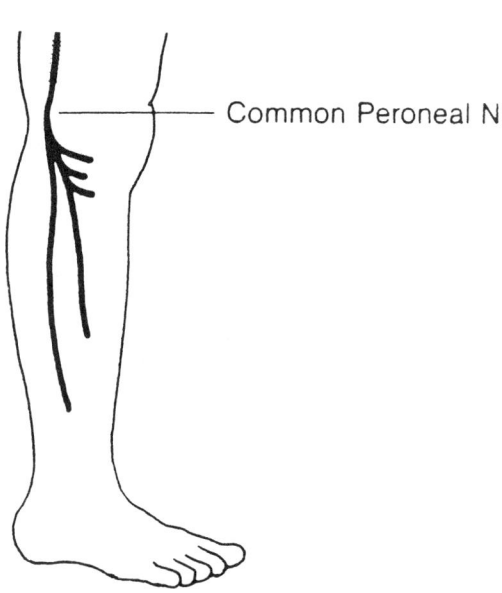

Figure 23-26

Most effective compliance results from impact penetration along the nerve pathway via either a baton strike or kicking technique. (See Fig. 23-27, Fig. 23-28, and Fig. 23-29)

Figure 23-27

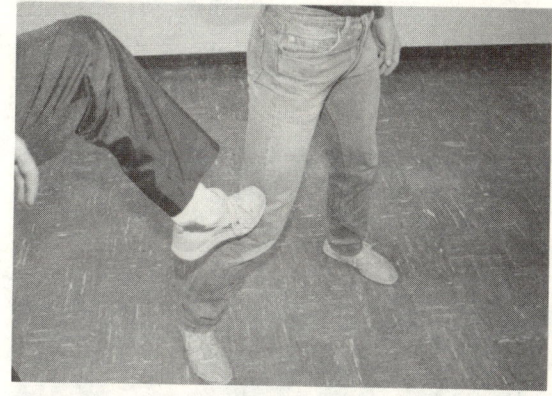
Figure 23-28

Figure 23-29

Tibial Nerve Area

The tibial nerve branches off the sciatic nerve and flows behind the knee along the back calf. (See Fig. 23-30)

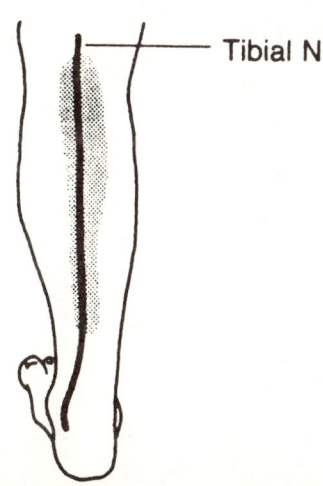
Figure 23-30

Most effective compliance is achieved when impact penetration occurs along the nerve pathway via either a baton strike or kicking technique. (See Fig. 23-31 and Fig. 23-32)

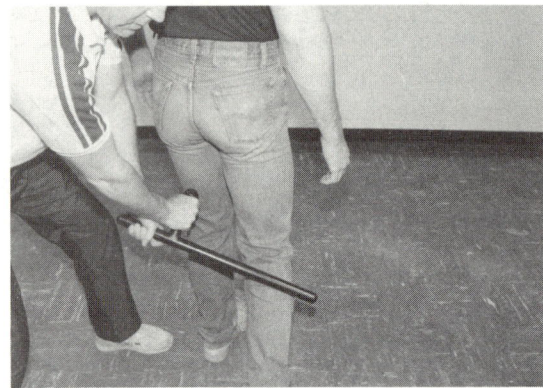

Figure 23-31

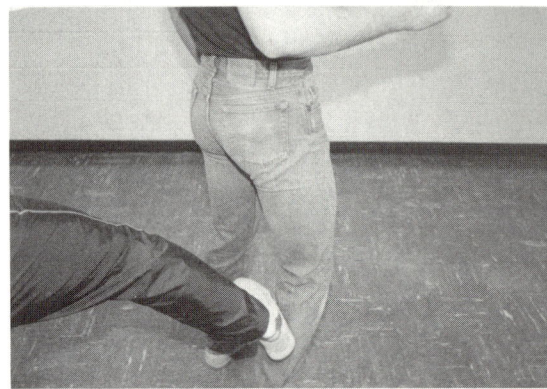

Figure 23-32

In review, it should be realized that nerve compression tactics can achieve several positive control goals, including:

1. Assist in establishing control over an aggressive individual.

2. Assist in the initiation and establishment of control holds.

3. Assist in reducing levels of resistance allowing for escape releases to be completed.

4. Assist in the enhancement of other non-lethal weapons available to the officer.

Chapter 24

INDIVIDUALIZED CHEMICAL IRRITANTS

Another force alternative available to the individual officer in appropriate situations is the aerosol projector containing chemical irritants. For over thirty years these weapons have proven their effectiveness in a variety of one-to-one confrontations. They allow the trained officer to selectively debilitate aggressive individuals from a distanced position of safety. The generic term "tear gas" has been used in reference to the two major component chemical solutions CN (Chloroacetophenone) and CS (Orthoclorobenzalmalonitrilc) which have been developed and marketed since the end of World War II. Today by and large the CN formulation is the primary product agent due to its rapid vaporization quality and its dispersal capacity. It is used most frequently for individual aerosol tear gas projectors because of its more immediate subject debilitation. CN is an organic compound and in its pure form, is a white crystalline solid resembling table salt. This agent is suspended in the delivery solvent and vaporizes on contact.

Contemporary marketing of these personalized aerosol chemical irritants is in seamless aluminum canisters, varying from 4" to 6" in length, and 1" to 1½" in diameter. Generally these units weigh between 4 oz. and 8 oz. Most units provide for 15 to 25 one second bursts and have an effective range of from 8 to 15 feet. It is recommended that to assure maximized effectiveness the units should be replaced once 50% or more of the formulation has been utilized or the units have been on the street for more than two years. The units should be checked monthly to insure that no leakage or damage has occurred. Lint, dirt, and other debris should be removed from the nozzle and trigger areas. Additionally, the officer may wish to trigger the unit once a month to guarantee its operation and to re-establish a perspective of targeting distance and trajectory path.

Although one of the weapon's greatest traits is that of simplicity, its tactical utilization does require familiarization and practice with the unit to understand its basic operational qualities and application effects. Throughout our explanation and demonstration of the weapon we will be using the most popular and proven product line, Smith and Wesson brand, Chemical Mace. According to its manufacturer Chemical Mace has had thousands of instances of street application with no long term debilitating results. Although the product line of Smith and Wesson Chemical Mace is quite extensive, the MK VI model of Chemical Mace will be illustrated throughout this lesson.

On the street this unit is holstered on the officer's non-weapon side. This unit's unique configuration assures rapid acquisition and positive, potential targeting. (See Fig. 24-1 and Fig.24-2)

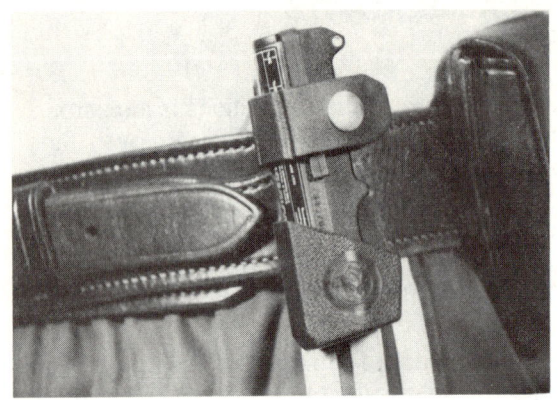
Figure 24-1
Holstered Unit on Belt

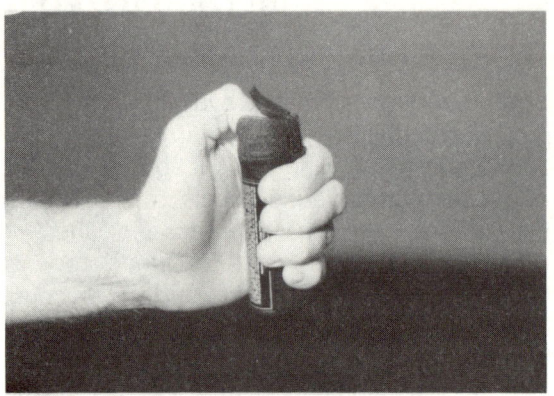
Figure 24-2
MK-VI Unit in Hand

Actual initiation of the chemical irritant during a confrontation must be based upon a reasonable assessment by the officer that the weapon would be legally proper and tactically feasible. Initially the officer should draw the weapon with his weak hand and hold it in a displayed position close to the officer's body to minimize takeaway opportunities available to an aggressive subject. Additionally, from this drawn and displayed position it may provide for deterent influence upon the subject and thus negate its actual full deployment. (See Fig. 24-3)

Figure 24-3
Officer in Drawn/Display Position

Once the decision has been made the weapon should be deployed by the officer via a short, one second burst into the chin area of the subject. From this body target site the high vapor pressure allows the CN formulation to travel up to the lacrinal glands around the eyes. (See Fig. 24-4)

Figure 24-4

Spray Impacting Chin area

Generally the CN formulation incapacitates in 1-5 seconds and remains effective in its debilitation qualities from 8-15 minutes. During this period the subject will usually experience temporary reflex closure of the eyes, excessive tearing, with intense burning of the skin and upper respiratory region. (See Fig. 24-5)

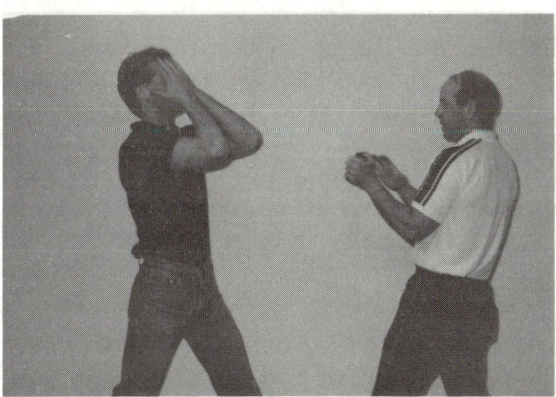

Figure 24-5

Subject Debilitation Response

The actual spraying of the irritant should be no closer to the face of the individual than 5 feet or no further than 15 feet to be most effective.

Obviously, no tactic or weapon is invincible nor are all cases of behavior totally predictable. Therefore, the officer must be prepared for a variety of technique shifts. For instance, if dispersal of the chemical formulation does not debilitate on initial application, the officer should commence another targeted one-second burst at the aggressor's chin. Especially at this point, but even during normal circumstances of debilitation, the officer should be prepared to neutralize the subject with an acceptable control tactic and eventual system of restraint. (See Fig. 24-6, Fig. 24-7, and Fig. 24-8)

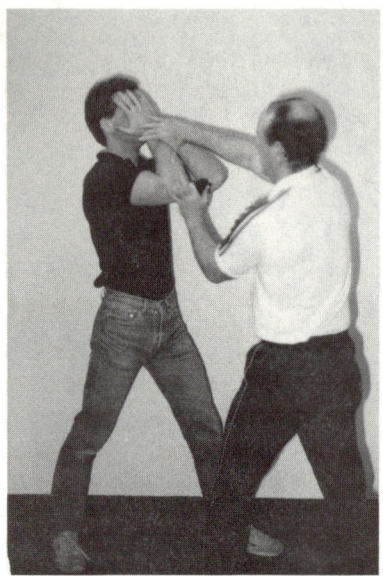

Figure 24-6

Initial Hand Grab to Begin the Control Tactic (Form 1)

Figure 24-7

Continuation of Takedown (Form 1)

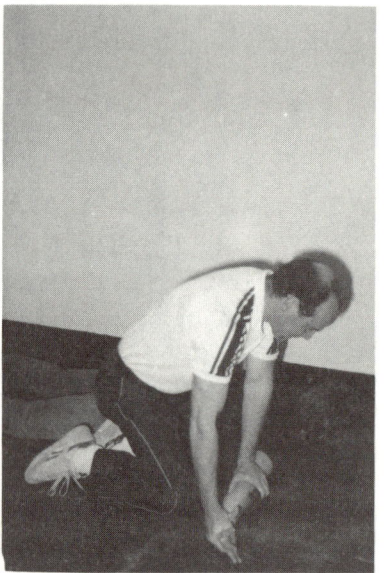

Figure 24-8

Prone Position/Pin Out

Under most circumstances the officer should effect his control options immediately after the lacrimating agents impact on the subject. Commonly, the subject will experience intense stinging and profuse tearing within seconds after application. He may experience choking and disorientation as well as increased dizzyness and balance loss. Under these conditions the officer has a more docile and confused individual prime for the officer initiating a variety of control tactics. If the subject becomes more cooperative the officer can reassure him that the effects are temporary and relief is forthcoming.

Obviously, not all human reactions or tolerances can be predicted. An aggressor may be under the influence of drugs, possess an extremely high tolerance for chemical irritation, etc. and be relatively unaffected with the use of the device other than primary visual impairment. Under these circumstances the officer must select an escalated weapon or tactical alternative.

It should be noted that as with other control tactics, communication with assisting officers is critical. In order to prevent a "chemical cross fire" leading to officer contamination the initiating officer should cue assisting officers that Chemical Mace is about to be used. This message can be in the form of code words or abbreviation so as not to pre-warn the resistant subject.

It should be noted that once the chemical irritant has been used, certain first aid provisions should be granted to the subject when safety permits. Allow the subject to flush his eyes with water. If still in the field, have the subject face toward an available wind to allow for the chemical vapors to escape. When possible remove the contaminated clothing from the subject to prevent more chemical vapors from being emitted. If affected areas appear to require additional medical attention, it should be provided as soon as possible.

Additionally, if Chemical Mace has been used indoors, contaminating exposed surfaces (coutners, pans, etc.) or substances (foods, etc.) they should be washed if appropriate or completely discarded.